# Studies in Brain and Mind

Volume 10

More information about this series at http://www.springer.com/series/6540

Alessio Plebe • Vivian M. De La Cruz

# Neurosemantics

## Neural Processes and the Construction of Linguistic Meaning

Springer

Alessio Plebe
Department of Cognitive Science
University of Messina
Catania, Italy

Vivian M. De La Cruz
Department of Educational Sciences
University of Catania
Catania, Italy

ISSN 1573-4536 ISSN 2468-399X (electronic)
Studies in Brain and Mind
ISBN 978-3-319-28550-4 ISBN 978-3-319-28552-8 (eBook)
DOI 10.1007/978-3-319-28552-8

Library of Congress Control Number: 2016932895



Printed on acid-free paper

This Springer imprint is published by Springer Nature
The registered company is Springer International Publishing AG Switzerland

# Contents

# Chapter 1
# Introduction

**Abstract** Neurosemantics is not yet a common term and in current neuroscience and philosophy it is used with two different sorts of objectives. One deals with the meaning of the electrical and the chemical activities going on in neural circuits. This way of using the term regards the project of explaining linguistic meaning in terms of the computations done by the brain. This book explores this second sense of neurosemantics, but in doing so, it will address much of the first as well, for we believe that the capacity of neural circuits to support linguistic meaning, hinges on their peculiar role in coding entities and facts of the world. It is an enterprise at the edge of the available state-of-the-art knowledge in neuroscience and specifically, in the growing understanding of brain computational mechanisms. We conceive neurosemantics, however, as the natural evolution of a long standing project that began in the early days of Boole's logic, the idea that semantics can be construed and explained in mathematical terms. Classical formal semantics, for a very long time, excluded from the analysis of language any account of mental processes, which on the contrary, became the central focus during the cognitive turn. Cognitive semantics, however, failed to provide a rigorous mathematical framework for semantic processes. Today, it is possible to begin explaining language by way of a new mathematical foundation, one that is empirically grounded in how the brain computes: neurosemantics. The way this book intends to contribute is twofold. One is to present a series of existing examples of neurosemantics in practice: early models addressing aspects of linguistic semantics purely in neurocomputational terms. The other is to try to identify the principles upon which models of this kind can be constructed, and their corresponding neural bases.

Neurosemantics is a word missing in the Oxford Dictionary for good reasons, it is not yet a common term, neither in neuroscience nor in philosophy, and its meaning is not yet well defined. The term was coined by Alfred Korzybski (1933), in his introduction to the second edition of *Science and Sanity*. Neologisms abound in his huge and highly controversial book of speculations on neurology and psychiatric therapy. Another term, that has met with better fortune is "neurolinguistic", borrowed by Gregory Bateson, and revived by Richard Bandler and John Grinder in the neurolinguistic programming (NLP) therapy. A legacy of Korzybski's neurosemantics still survives today in the International Society of Neuro-Semantics lead by Michael



A. Plebe, V.M. De La Cruz, *Neurosemantics*,
Studies in Brain and Mind 10, DOI 10.1007/978-3-319-28552-8_1

Hall, a negligible competitor of NLP in psychotherapy. Korzybski already used the word "semantics" without "neuro" so broadly that it became almost meaningless, with no way of establishing a comprehensible concept by their conjunction. In this book, the only concern with neurosemantics as used by Korzybski, is strictly limited to this brief historical note.

In more recent times, and in more scientific contexts, neurosemantics has been used with two different sorts of objectives. One deals with the meaning of the electrical and the chemical activities going on in neural circuits. This is, for example the usage of neurosemantics by Churchland (2001), Ryder (2004), or Breidbach (2007). The second, deals with the same type of semantics studied for years in philosophy: the meaning of language, but trying to offer an explanation in terms of the neural computations performed when people listen to and understand utterances. This is the main usage of neurosemantics, for example, in Pulvermüller (2012), and the project pursued there is not much different from that of Feldman (2006), who instead does not explicitly use the term.

This book explores this second sense of neurosemantics, but in doing so, it will address much of the first meaning as well, for we believe that the capacity of neural circuits in humans to support linguistic meaning, hinges on their peculiar role, shared by many other animals, of coding entities and facts of the world. Even if combining sequences of sounds for conveying meaning through words and sentences is a strategy exploited by humans only, its realization by way of neural circuits appears to be rooted on their disposition to code for things in the world, something that is shared in a wide variety of forms, with other animals. This continuity is attested by famous cases of non human animals being successfully trained to use simplified or partial forms of language (Savage-Rumbaugh et al. 1998; Pepperberg 1999).

A different type of continuity is followed in this book, between the sketch given of the new field of neurosemantics, and that of classical semantics. The idea is that semantics can be captured and explained, by a sort of computation. Formal logic has been the extraordinary attempt to mathematicize the composition of meaning by words. Set forth by Boole (1854) borrowing from algebra, it took its contemporary form at about the beginning of the last century thanks to the work of Frege (1881), Russell (1903), and Wittgenstein (1922), among others.

One well known posit in logic was antipsychologism, the view that language should be analyzed as an abstract entity, in isolation, separated from the minds that comprehend and use it. Any attempt to shift the analysis inside mental mechanisms was considered as misleading. While this position had historical motivations and epistemological merits, it became increasingly unsatisfactory, as the new science of the mind progressed rapidly in the second half of the past century. Inside the newly born field of cognitive science, investigation of the mind became the central focus of a joint effort between philosophers, computer scientists, psychologists and linguists. Cognitive semantics has been the resulting enterprise, one that has brought semantics back inside the mind, a project that ranges from Rosch's (1978) prototype theory, to the radial categories of Lakoff (1987), to the cognitive grammar of Langacker (1987), and includes many other theories.

The cognitive turn not only shifted the investigation of semantics towards the mind, in doing so, it also fostered the exploration of aspects of meaning that until then had been considered only marginally. The privileged building blocks in the field of semantics have been propositions, and the greatest efforts have concerned the study of truth conditions of complex sentences. Not much attention had been dedicated to the analysis of single word meanings, except for functional terms, like conjunctions or quantifiers. A new fervor of research on lexical meaning naturally espoused the psychological search of mental concepts and categories.

While cognitive semantics led to many important innovations such as those mentioned above, it failed to provide a new mathematical framework comparable to those furnished by logic. This was certainly not due to a generally skeptical attitude towards a computational approach, on the contrary. Part of the manifesto of cognitive science was the adoption of "Computational Theory of Mind", the philosophical view that the basic activity of the mind is that of performing computations over mental states, and that the essential job of cognitive scientists was that of identifying the abstract functions we all compute, this task however, has been scarcely practiced within cognitive semantics. In a way, the worries of analytic philosophy against psychologism were founded: it is difficult to maintain scientific rigor when adventuring across the meanders of the mind. Cognitive semantics is rich in deep intuitions, but is also characterized by an extreme vagueness, and often a naiveté, regarding questions of mathematical modeling, as Seuren (2004) has remarked.

In our opinion, the main cause of the limited effect of the cognitive turn in laying down the foundations of a new semantics, is in the gap between the proposed models, and the empirical evidence of their correspondence with brain processes. The main concepts in cognitive semantics are inventions of their proposers, without any reference to the possible computational correlates in the brain.

The situation has changed dramatically in the last half century, during which much light has been shed on the kind of computation carried out by neural circuits. The enterprise of setting up the mathematical grounds for modeling neural behavior is ongoing, in the growing domain of neural computation (Sejnowski et al. 1988), today included in the broader area of theoretical neuroscience (Dayan and Abbott 2001). In the meantime, the expanding body of knowledge provided by neuroscience has gained a philosophical foundation, under the umbrella of "neurophilosophy" (Churchland 1986). Its main assumption is that mental activity is brain activity, and as such, can be subject to scientific methods of investigation, properly guided by cognitive science in characterizing the repertoire of phenomena to be explained. The Golden Fleece in scientific investigation is computational modeling. It clearly needs to build on other methods, such as empirical neurocognitive studies, but it has the unique privilege of offering explanations in the rigorous language of mathematics. There are different opinions on what exactly a neurocomputational explanation is. A recent influential position is that the explicatory power of a neurocomputational model depends on how its structure maps well with brain *mechanisms*, in the technical sense of system components and functions performing the studied behavior (Piccinini 2006; Kaplan 2011). According to this criterion, the virtues of a neurocomputational model is not only in its ability to predict a behavior, but also in how accurately its main mathematical terms correspond to constituents of brain processes.

This degree of accuracy has its upper limit in the knowledge of the neurophysiological mechanisms, which in the case of cognitive functions such as language are dense as well as scattered, but in continuous progress. So, the time is ripe for turning semantics into neurosemantics. The way this book intends to contribute is twofold. One is to present a series of existing examples of neurosemantics in practice: early models addressing aspects of linguistic semantics purely in neurocomputational terms. The other is to try to identify the principles upon which models of this kind can be constructed, something for which mathematics is available, and the underlying neural bases. Here is where the book will meet the other sense of neurosemantics: how neural circuits can carry representations of the world. It is a widely debated field, which includes positions that simply deny the notion of representation (van Gelder 1998). Even without a philosophically defendable definition of representation in hand, we will continue to refer to neural representation, as the most useful pragmatic concept in describing how brain circuits capture knowledge of the external world. We are not so alone in this enterprise (Bechtel 2014), and our sketch of representational mechanisms will privilege those we deem essential to the semantics of language.

The contribution to neurosemantics, is mainly a collection of models developed by the authors, where certain specific aspects of the semantics of language are reproduced, based on an essential reconstruction of relevant cortical and subcortical areas, each simulated by a mathematical formulation respectful of the main biological neural processes. These models are presented in the second part of the book, and discussed as neurosemantic practice. Some of these models target one of the core problems of semantics, the reference of nouns, and in particular of nouns with a strong perceptual characterization. Others address the semantics of predicates, with a detailed analysis of color attributes. This domain has particular relevance in philosophy of language, as a testbed for the hypothesis of language relativism. This debate has long been dominated by the view that world languages follow a universal scheme in naming their basic colors (Berlin and Kay 1969), until the recent discovery by Roberson et al. (2004) of two counterexamples. Himba, spoken in Northern Namibia, and Berinmo, spoken in Papua New Guinea, have their own linguistic color categories organized not only differently from each other, but also differently from the universals scheme. The emergence of their color term semantics has been simulated in the neurocomputational model presented in the book, and compared with the English one.

The search of the neural roots of meaning can also contribute to clarifying the semantics of certain classes of words that have puzzled traditional semantics for decades, without any agreed upon solution. This is the case of moral sentences, which defeat the standard semantic analysis of truth conditions. Several solutions have been proposed, such as expressivism (Gibbard 1990), the idea that a sentence like "stealing is wrong" merely express an attitude of disapprobation against stealing, and therefore eludes truth conditional semantics. This, like other solutions, engender a set of logical problems, the most severe being the Frege-Geach (1965) embedding problem. Since every moral judgment can be embedded in logical constructs like conditionals, for example in "if stealing is wrong then you should not teach it", it is unclear how they can work at the same time as truth-apt and non-truth-apt components. We will show how neurocomputational models of affective and decision brain areas are starting to provide a coherent picture of the emergence of moral meaning.

Neurosemantics has become a reality, and many aspects of language, not covered by the models developed by the authors, are currently being explored by other scholars. A general picture will be provided, together with considerations on the future trends of this challenging enterprise.

## References


Bechtel, W. (2014). Investigating neural representations: The tale of place cells. *Synthese*, 1–35. doi:10.1007/s11229-014-0480-8.

Berlin, B., & Kay, P. (1969). *Basic color terms. Their universality and evolution*. Berkeley: California University Press.

Boole, G. (1854). *An investigation of the laws of thought, on which are founded the mathematical theories of logic and probabilities*. London: Walton and Maberley.

Breidbach, O. (2007). Neurosemantics, neurons and system theory. *Theory Bioscience, 126*, 23–33.

Churchland, P. S. (1986). *Neurophilosophy*. Cambridge: MIT.

Churchland, P. M. (2001). Neurosemantics: On the mappings of minds and the portrayal of worlds. In K. White (Ed.), *The emergence of mind*. Milan: Fondazione Carlo Erba.

Dayan, P., & Abbott, L. F. (2001). *Theoretical neuroscience*. Cambridge: MIT.

Feldman, J. A. (2006). *From molecule to metaphor: A neural theory of language*. Cambridge: MIT.

Frege, G. (1881). Über den Zweck der Begriffsschrift. Sitzungsberichte der Jenaischen Gesellschaft für Medizin und Naturwissenschaft XV. Reprinted in Angelelli, I. (Ed.). (1964). Begriffsschrift und andere Aufsätze. Hildesheim: Olms.

Geach, P. T. (1965). Assertion. *The Philosophical Review, 74*, 449–465.

Gibbard, A. (1990). *Wise choices, apt feelings – A theory of normative judgment*. Cambridge: Harvard University Press.

Kaplan, D. M. (2011). Explanation and description in computational neuroscience. *Synthese, 183*, 339–373.

Korzybski, A. (1933). *Science and sanity – An introduction to non-Aristotelean systems and general semantics* (2nd ed., 1941). New York: Institute of General Semantics.

Lakoff, G. (1987). *Women, fire and dangerous things. What categories reveal about the mind*. Chicago: Chicago University Press.

Langacker, R. W. (1987). *Foundations of cognitive grammar*. Stanford: Stanford University Press.

Pepperberg, I. M. (1999). *The alex studies: Cognitive and communicative abilities of grey parrots*. Cambridge: Harvard University Press.

Piccinini, G. (2006). Computational explanation in neuroscience. *Synthese, 153*, 343–353.

Pulvermüller, F. (2012). Meaning and the brain: The neurosemantics of referential, interactive, and combinatorial knowledge. *Journal of Neurolinguistics, 25*, 423–459.

Roberson, D., Davidoff, J., Davies, I. R., & Shapiro, L. R. (2004). The development of color categories in two languages: a longitudinal study. *Journal of Experimental Psychology: General, 133*, 554–571.

Rosch's, E. (1978). Principles of categorization. In E. Rosch & B. Lloyd (Eds.), *Cognition and categorization*. Mahwah: Lawrence Erlbaum Associates.

Russell, B. (1903). *Principles of mathematics*. Chicago: Chicago University Press.
Ryder, D. (2004). SINBAD neurosemantics: A theory of mental representation. *Minds and Machines, 19*, 211–240.
Savage-Rumbaugh, S., Shanker, S., & Taylor, T. (1998). *Apes, language and the human mind.* Oxford: Oxford University Press.
Sejnowski, T. J., Koch, C., & Churchland, P. S. (1988). Computational neuroscience. *Science, 241*, 1299.
Seuren, P. (2004). How the cognitive revolution passed linguistics. Language and revolution: Language and time 89(63–77).
van Gelder, T. (1998). The dynamical hypothesis in cognitive science. *Behavioral and Brain Science, 21*, 615–665.
Wittgenstein, L. (1922). Tractatus Logico-Philosophicus. Trench, Trubner & Co., London, trad. it. di Amedeo G. Conte 1964.

# Part I
# The Meaning of Neurons

This first part sets the stage for the neurosemantic enterprise undertaken in this book, the grounding of language meaning in the brain. In doing so, it will also address, in the second part, different conceptions of neurosemantics: how neurons, with their axons, dendrites, ion channels, convey meaning. These are the same ingredients that in humans, allow them to augment the representation of the word through fine-grained chunks of sound. There is nothing more in our brain with respect to non-speaking animals. Therefore, much of what is expected from a neural semantics of language should rely on more general properties of the neural system.

Most likely every bit of the huge body of current neuroscientific knowledge plays a role in how the brain makes sense of the world. Aware of its leaving out a great deal, this part is nothing like a broad introduction to basic neuroscience. It merely focuses on describing a limited number of elements and mechanisms that are considered to contribute in an essential way to the computational construction of meaning in the neural system. As stated in the introduction, privileging the computational approach has a double valence in this book. On one hand, it is the continuation of the idea that the semantics of language can be captured and explained in mathematical terms, a concept that has been a cornerstone in the field of logic. On the other, it is rooted in the evidence that the brain itself computes, and this is the main aspect covered in this part. It is an approach that is much in line with what Shagrir (2010) has called the "San Diego style" of computation, with reference to the mix of philosophers, neuroscientists, and computer scientists, working together in universities along the western seaboard of the United States these last couple of decades. This flavor of computation departs from the standard paradigm of the execution of a digital program, and moves towards mathematical frameworks proper of neurocomputation, while still preserving the concept of representation.

What it exactly means "to compute", is a wide open debate, which has flourished distinctively with respect to brain activity (Piccinini and Scarantino 2010; Piccinini and Bahar 2013; Fresco 2014; Piccinini 2015), and we will discuss how the account of neural computation embraced here fares within this debate.

The three chapters of this first part can be conceived as a progression from neurophysiology to mathematics. The first chapter introduces the elementary computational devices of the brain, and is of course mainly a celebration of the neuron, the cell with which nature greatly succeeded in elaborating electricity without the help of metals and semiconductors. From the properties of the neuron, and of large assemblies of neurons, it is possible to construct a number of mechanisms, which can explain how neurons represent, this is the content of the second chapter. The last chapter will take a further step, that will lead us from mechanisms to algorithmic principles, more precisely the small set of principles, which will be used in building the neurosemantic models that will be the content of this part.

## References

Fresco, N. (2014). *Physical Computation and Cognitive Science*. Berlin: Springer.

Piccinini, G. (2015). *Physical Computation: A Mechanistic Account*. Oxford: Oxford University Press.

Piccinini, G., & Bahar, S. (2013). Neural computation and the computational theory of cognition. *Cognitive Science, 34*, 453–488.

Piccinini, G., & Scarantino, A. (2010). Computation vs. information processing: Why their difference matters to cognitive science. *Studies in History and Philosophy of Science, 41*, 237–246.

Shagrir, O. (2010). Computation, San Diego style. *Philosophy of Science, 77*, 862–874.

# Chapter 2
# The Computational Units of the Brain

**Abstract** Every mathematical framework is developed around some basic computational item, for example, sets, numbers, and vectors. This is the starting point for a computational view of the brain as well, the basic units have to be specified. While in the ethereal world of mathematics the basic components can be arbitrarily assumed, even invented from scratch, the case of the brain is constrained by its biophysical structure. The current view today is dominated by the paradigm constructed by Ramón y Cajal, where the neuronal cell is the basic computational device of the brain. Enormous progress has been achieved in characterizing the computational properties of the brain under this paradigm, which will partly be reviewed in this chapter, limiting ourselves to those aspects that are useful for understanding the representational power of the brain. However, like any scientific paradigm, the so called "neuron dogma" might possibly change in the future, there are scholars for example (London and Häusser, Annu Rev Neurosci 28:503–5032, 2005) that argue that dendrites display autonomous computational capabilities, and might be a better candidate for the title of basic computational unit.

Despite many attempts, it is still difficult to spell out what the most specific function of the neuron as a computational device is, and how that makes it so different from other man-made computational devices. We favor an idea advanced by Turing (1948, Intelligent machinery. Technical report, National Physical Laboratory, London, reprinted in Ince DC (ed) Collected Works of A. M. Turing: Mechanical Intelligence, Edinburgh University Press, 1969) long ago: neurons have no special built in function, but that of being able to learn virtually any function, by experience. Plasticity is the term in neuroscience that includes the biological mechanisms that explain how neurons work as extraordinary learning machines.

The last section of this chapter deals with a special organization of neurons, that has long been held to deserve a specific computational description (Stevens, What form should a cortical theory take? In: Koch C, Davis J (eds) Large-scale neuronal theories of the brain. MIT, Cambridge, pp 239–255, 1994), and seems to be the privileged site of semantic representations: the cortex.

 
A. Plebe, V.M. De La Cruz, *Neurosemantics*,
Studies in Brain and Mind 10, DOI 10.1007/978-3-319-28552-8_2

## 2.1 The Neuron

The neuron is the fundamental cell of the nervous system, brain included. It is the key element that sets animals apart from all other living organisms, offering them the inestimable advantage of voluntary motion, allowing the search for food, and the escaping of risks. Being based on electricity, it required nature to invent a way to deal with it. Man made electrical power is conducted by metals, like copper, the fastest available conductor. Computers, the artifacts managing electricity at a level of sophistication comparable to the brain, are made of semiconductors, such as silicon and germanium. Nature opted for the only electrical conductors compatible with organic materials: ions.

The biophysical breakthrough of exploiting electric power in animals has been the *ion channel*, a sort of natural electrical device, whose details have been discovered only recently (Neher and Sakmann 1976). It allows the flow of a specific type of ion only, across a cellular membrane, and only under certain exclusive circumstances, typically being the difference in voltage between the internal and the external areas of the cell. The first ion channel to appear in evolution was the potassium $K^+$ channel, which appeared about three billion years ago in bacteria. It evolved into the calcium $Ca^{++}$-permeable channel in eukariotes, and finally into the sodium $Na^+$ channel, already found 650 million years ago in both ctenophora (kind of jellies) and early bilateria (animals with bilateral symmetry) (Zakon 2012). From then on sodium channels developed in all animals, and is currently the most important neural channel. Its success is explained by the abundant availability of sodium in the marine environment.

This history is confirmed by comparisons between extant species with non $Na^+$ $Ca^{++}$ only ion channels, such as fungi and animals with simple nervous systems, such as ctenophora and sea anemone (Liebeskind et al. 2011), but many, if not most, of the details are still uncertain, and inextricably linked to a better understanding of the phylogeny of metazoans. Moroz (2009) proposed a hypothesis for the phylogeny of the neuron independent from the history of ion channels. One shared prerequisite of all neurons, from a genomic standpoint, is the capacity to express many more genes and gene products than other cell types. In fact, other cells can also exhibit massive gene expression, as a result of severe stress responses, and typically before death. Neurons might have evolved in ancestral metazoans from other types of cells, as the result of development in the adaptive response to localized injury and stress, which gradually stabilized in cells supporting and maintaining the expression of multiple genes and gene products in normal conditions.

It may seem that in the search to understand how the brain supports semantics, going back to the dawn of animal life is far too long a path to travel. Let us just mention that recent research is trying to establish the first glimmers of intelligent behavior thanks to electricity, going even further back. Traces of neural-like electrical communication has even been found in plants (Baluška et al. 2005). Not only has a discipline named *plant neurobiology* recently emerged (Brenner et al. 2006), but even the concept of *plant cognition* is currently being discussed (Garzón and Keijzer 2011). Perhaps plant cognition will reveal itself to be not as extravagant as is expected, but it certainly cannot involve language. After this brief journey discussing the origins of the neuron, we can now proceed in describing it as the chief device found in linguistic animals.

### *2.1.1 Cajal's Doctrine*

Contrary to the natural history of the neuron, with its beginnings so far away in evolutionary time, its scientific history is surprisingly recent. There was absolutely no clue to how the brain worked until the end of the nineteenth century: no idea that it was composed of neural cells, no idea that it functioned as a complex electrical engine. Sigmund Freud in his early scientific career did some investigations on the nervous system (Freud 1885) leading him later to speculate about the exchange of "energy" between the cell connections (Freud 1895). According to Sulloway (1982) this idea proved to be fecund, inspiring other themes found in his famous theory of psychoanalysis, such as the concept of the redirection of psychic energy. He was not able to provide details, however, on the nature of this "energy". Decades before, a Scottish philosopher, Bain (1873), theorized that communication in the nervous system was based on weak electric currents. After the formulation of cell theory by Theodor Schwann (1839), stating that all organisms are composed of individual cells, several biologists, such as Wilhelm His, Auguste Forel, and Fridtjof Nansen, argued that the same idea could apply to the nervous system as well. However, the prevalent view at the time was that the nervous system was an exception to cell theory, because of its being organized in a reticular way, a theory boosted by the misinterpretation of Deiters (1865) of axons as emerging from dendrites, and having in Gerlach (1871) its major proponent. The reticularist view was dismissed by Santiago Ramón y Cajal, one of the founders of contemporary neuroscience. He was a scientist endowed with an extraordinary capacity to observe and understand brain structure. In his master work *Textura del sistema nervioso del hombre y de los vertebrados* (Ramón y Cajal 1899) he stated, on the basis of hundreds of empirical observations, and carefully reasoned arguments, that the cell theory fully applied to the brain. He maintained for the main cell of the nervous system the name "neuron" given by Waldeyer-Hartz (1891), a German anatomist who had a fortunate lexical creativity, having also coined the term "chromosome". What is now known as the *neural doctrine* was born.

In being an instance of the cell theory applied to the nervous system, of course the main tenet of the neural doctrine is that there is a specific cell, the neuron, that is separate and distinct and not anatomically continuous to other neuronal cells, and constitutes the structural and functional unit of the nervous system. In addition, the theory establishes a main division of the neuron in parts: the dendrites, the soma, and the axon. Cajal kept the names introduced by Kölliker (1893), these elements are shown schematically in Fig. 2.1. The axon has several terminal arborizations, which make close contact with dendrites or the soma of other neurons, included in

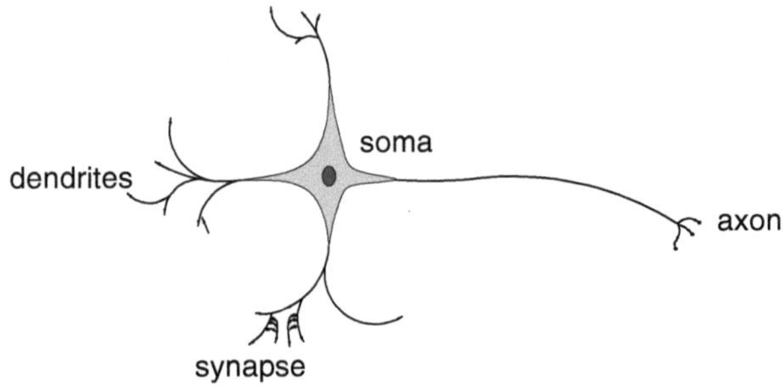


**Fig. 2.1** Simplified scheme of a neuron

the neural doctrine is also the concept of *functional polarity*, the assumption that the flow of information between two cells is in one direction only: from the axon to the dendrites.

Soon after, the picture was completed by another founding father of neuroscience, Charles Sherrington, who introduced the concept of the *synapse*, a tiny space between a transmitting axon and a receiving dendrite. In his Croonian lecture in 1897, he theoretically postulated the need for such a communication nexus (Sherrington 1941). As we will report in Sect. 2.2.1, it was impossible to investigate such small spaces at the time, and so Sherrington's postulation was based on purely theoretical grounds.

### 2.1.2 Ions and Computation

The path opened by Cajal and Sherrington, among others, was soon traveled by a growing community of scientists, who discovered an increasing amount of information on the neuron. Initially, research efforts did not include a commitment to analyzing the neuron as a computational device. This particular characterization of the neuron was basically impossible during the first decades of the last century, considering that the use of the notion of "computation" in this domain did not yet exist. Way before the birth of computers, Max Nordau (1895) had shown foresight in using the fortunate expression "true computing machines", with reference to the eye and the ear, as specific collectors of ondulatory energies. After Turing (1936), mainstream research on the neuron was scarcely interested in this issue, with the exception of McCulloch and Pitts (1943) whose contribution will be discussed in Sect. 5.1, and of course Turing (1948) himself. For what it precisely means for a neuron to compute, in light of the current philosophy of computation, we defer to Sect. 3.1.4, an informal notion of computation as signal manipulations describable in mathematical terms suffices in following the pioneering characterization of the neuron in this sense.

At that time, it was clear that the function of neurons was essentially electrical, and thus, the search for a quantitative description of their electronic properties was necessary in order to fully understand them. Important support in this direction came

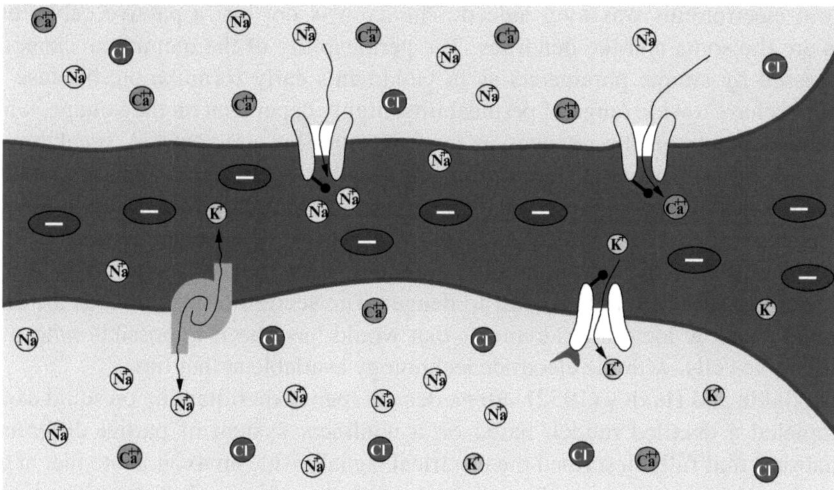


**Fig. 2.2** Detail of a dendrite, with some of the main ions contributing to its electrical potential: Sodium $Na^+$ Potassium $K^+$ Calcium $Ca^{++}$ Chloride $Cl^-$. The ions marked with "–" are large organic ions, that cannot travel the membrane, and contribute to the negative potential at rest

by way of the study of man made electricity, because of the extensive mathematical methods developed early in the twentieth century. At the time, a very particular business needed to be handled, that of the transatlantic telegraph line, laid in the early 1850s. Lord Kelvin (1855) himself engaged in formulating the equation describing the variation of voltage along the cable, a task that was not too difficult for him, thanks to the analogy with the problem of heat conduction, a topic he knew quite well. Axons, in the end, are cylinders much like cables, and Hoorweg (1889) first recognized that much of the mathematical treatment of electric cables could be used for describing the nervous electrotonus. As noted in the beginning of this section, a significant difference, however, is that while in cables the carriers are free electrons, inside the brain they are ions.

In Fig. 2.2 the principal ions involved in the neural potential are shown. As said before, sodium ions dominate the scene in the brain. There is a typical differential distribution of ions inside and outside the neuron, with higher concentrations of $K^+$ inside, and $Na^+$, $Ca^{++}$ and $Cl^-$ concentrated more in the extracellular space. A characteristic of almost all neurons is an internal negative potential at rest, typically of −40 to −90 mV, due to an excess of negative charges with respect to the outside of the cell. This is due to the presence of organic ions, too large to leak across the membrane, and because potassium-permeable channels allow a continual resting efflux of $K^+$.

A step forward towards an understanding of the electricity inside nerves was accomplished by Goldman (1943), who worked out an equation of the voltage across a membrane of a generic cell, in terms of ion concentrations and the permeability of the membrane. Still, the path to a mathematical description of the neural electrotonus was long indeed. The axon is not just a passive cable, even less are the soma and the dendrites. The permeability of the membrane cannot be expressed by simple parameters as in Goldman's early formulation, because ion channels have a wide range of permeability, highly dependent on the voltage, which in turn depends on the changes in ion concentration gradients. A breakthrough came in the mid twentieth century, made possible by an extravagance of nature. Certain types of squid are equipped with extraordinarily large axons, up to 800 μm of diameter, against the typical 1–2 μm of mammals. Their giant neurons activate the contraction of the mantle muscle, producing a jet propulsion effect that allows the squid to react remarkably fast to danger. The section of squid axons are large enough to allow local measurements that would have been impossible with most other nerve cells, with the electrode technology available at that time.

Hodgkin and Huxley (1952), after a decade spent experimenting on squid axons, completed a detailed model, based on a nonlinear system of partial differential equations, that fully described the electrical signal inside an axon. Note that at that time there was no direct evidence for the existence of ion-selective channels, and Hodgkin and Huxley speculatively formulated separate equations for the voltage-dependence of $Na^+$ and $K^+$ channels, based on their experiments.

One of the main achievements of the Hodgkin-Huxley model was the reproduction of the *action potential*, the characteristic electrical waveform, first described by du Bois-Reymond (1849), who named it *negative Schwankung*, today informally denoted as "spike". It is an abrupt electrical impulse, that propagates along the axon with an almost identical shape, shown in Fig. 2.3. The rapid rising of the voltage is an avalanche effect lead by $Na^+$. As soon as the internal potential reaches a certain threshold, the $Na^+$ conductance increases, causing $Na^+$ to enter the neuron, thus depolarizing the membrane potential, which triggers an even larger increase in the conductance of the channel. At a slower rate the depolarization activates the $K^+$ conductance as well, causing $K^+$ ions to leave the cell, repolarizing the membrane potential, which becomes briefly more negative than the normal resting potential.

Despite its venerable age, the Hodgkin-Huxley model is still the point of reference for the axon electrotonus. More recent developments have addressed

**Fig. 2.3** Typical waveform of the action potential

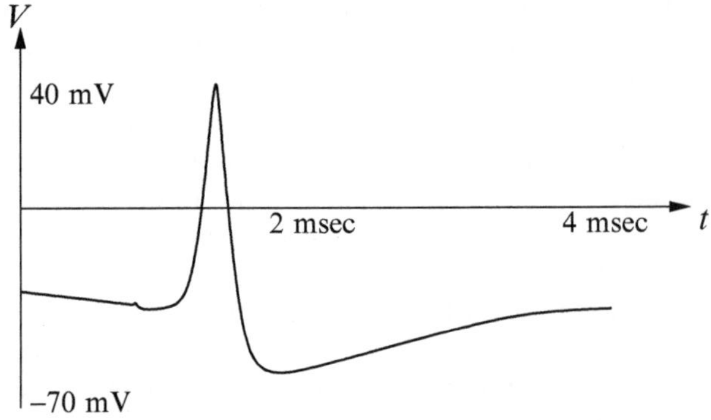

the contribution of other ions (Golowasch et al. 1992), the integration of detailed channel mechanisms (Rinzel 1990; Hille 1992), the extension to the soma (Bush and Douglas 1991), and dendrites (Traub et al. 1991). Not only did the Hodgkin-Huxley model open the way to a quantitative description of neural signaling, it has also played an influential role in the birth of computational neuroscience, and has been used as the paradigmatic example in the debate over the nature of explanation in neuroscience (Bogen 2005; Kaplan and Craver 2011). According to Craver (2007), the Hodgkin-Huxley model is a brilliant predictive model, but not a mechanistic model, and as such, does not give an explanation of the action potential: "In the HH model, commitments about underlying mechanisms are replaced by mathematical constructs that save the phenomena of the action potential much as Ptolemy's epicycles and deferents save the apparent motion of the planets through the night sky."

The Hodgkin-Huxley model describes one of the most important electrical phenomena in the neuron, the action potential, but it is still far from characterizing the neuron as a computational unit. First, the axon is just one component of the whole neuron, furthermore, the firing of a single neuron is meaningless in the context of the brain. As we will see in Sect. 4.1 a basic trait of brain computation is the cooperative interaction between a large number of neurons. For even the most elementary brain task, the activation or the resting of a single isolated neuron is irrelevant, while, for example, in computers, one single machine instruction, affects the whole program. Thus, a computational account of the neuron should, first and foremost, mathematically describe the interaction between more than one neuron. The neurophysical side of this issue is the theme of Sect. 2.2.1.

Let us add that the Hodgkin-Huxley model has inspired various types of modeling, oriented to being included in a network of interacting units, making simplifications on the details of the equations. One of the most common is the so-called *integrate-and-fire* model, where the potential is given by a single linear differential equation (Gerstner 1999):

$$\tau \frac{\partial V}{\partial t} + V(t) = kI(t) \tag{2.1}$$

where $V$ is the potential, $I$ is the current, and $\tau$ is a time decay constant, $k$ and equivalent electric resistance. The action potential is produced by adding the following condition:

$$V(t) > \theta \tag{2.2}$$

$$\frac{\partial V}{\partial t} > 0 \tag{2.3}$$

where $\theta$ is the threshold for the neuron to fire. Under certain assumptions this model can approximate the Hodgkin-Huxley model to 90 % (Kistler et al. 1997).

### 2.1.3 How the Doctrine Fares Today

For the purposes of this book, independently from the adequacy of description given by the Hodgkin-Huxley or equivalent models, it is important to confirm whether the neuron can still be held as the computational unit of the brain, as asserted by Cajal's neural doctrine. More than one century has passed, and a recurrent destiny for scientific doctrines, or paradigms if you like (Kuhn 1962), is that of being replaced by others. If we were to bet on which emerging paradigm might contend with the neural doctrine today, dendrites would probably have the best odds.

Not one of the rules constituting the neural doctrine has been immune to scrutiny in different moments in time (Bullock et al. 2005). Neurons have been found to be connected not only through conventional synapses, but also by more direct channels called gap junctions, that provide neurons with cytoplasmic continuity (Connors and Long 2004; Fukuda 2009), a posthumous consolation for reticularists. It has also been discovered, that sometimes, action potentials can travel backwards from the axon and soma regions into the dendrites (Waters et al. 2005). Recent research has pointed to a role for glial cells in brain computations, influencing axonal conduction and synaptic transmission (Fields and Stevens-Graham 2002). Remarkably, most of these aspects were already envisioned by Cajal, and nevertheless, none of the exceptions to the standard rules in the neural doctrine, just listed, point to a new and better candidate, other than the neuron as the computational unit of the brain.

Only dendrites may reasonably aspire to such an honor. Being much thinner than axons, studies exploring them followed much later those of axons. The biophysics of dendrites and their computational roles, became a focus of direct experimental research only in the late 1960s, mainly thanks to the work of Wilfrid Rall (1967). Today London and Häusser (2005) are among the main proposers of dendritic computation, they advance a speculative argument based on the disproportion between dendrites and axons in a neuron. Typically, a brain neuron provides just one output through its axon, based on thousands of synaptic inputs at its dendrites. This final conversion equates to a mathematical function projecting a huge space onto a narrow one. Such a complex function would be unrealizable by the simple summation of dendritic contributions in the soma potential, so dendrites should thus be warranted with much more computational power. Based on a series of empirical results, London and Häusser presume a series of computational abilities dendrites have, from basic logical operations to coincidence detection (see Sect. 3.2) and lowpass filtering. In a similar vein, Sidiropoulou et al. (2006) contend that the computation performed by a single neuron is far too complex for it to be used as a basic computational element. A discussion on the computational autonomy of dendrites with respect to whole neurons can be found in Cao (2011, 2014), under the perspective of semantic information.

Unlike the theory of Shannon and Weaver (1949), which focused on the reproduction of messages from a source to a destination, and the quantitative measurement of the information in a message, semantic information is concerned with what a particular message stands for or means (Dretske 1981). Inside the theory of semantic information, it is possible to conceive under which circumstances a message carries meaningful information, taking into account the sender of the message, the receiver, and the context of the communication. Some specific requirements to be satisfied by an entity, in order for it to qualify as a genuine sender or receiver of information, can be specified (Cao 2011, p. 58):

- Signals that differ only a little can result in dramatically different actions from a receiver. (e.g. "I love you" and "I loathe you" [...]).
- Significant differences in the physical features of a signal might make no difference to the receiver. (e.g. receiving a party invitation by mail vs. in person).
- The effects of signals will be strongly context-dependent and easily changed. (e.g. phone ringing when you expect good news vs. when you expect bad news).

Cao scrutinized how brain units stand up to these requirements. The answer is: poorly. The subunits of a neuron, for example, fail to meet the criteria of flexibility in responding to a signal on the basis of its meaning. The whole neuron itself, meets the requirements for being a receiver only in a limited way, it is compared by Cao to the man inside Searle's Chinese Room, who keeps taking inputs and producing outputs in total ignorance of their meaning and of the world outside. Better candidates in the brain as sender or receiver of information are, for Cao, groups of neurons, where the contribution of the firing of a single neuron becomes less important. How neurons group together in purposeful circuits will be widely discussed in the next chapter.

We must add that the neural doctrine continues to have a large number of supporters. Azmitia et al. (2002) even complain that missing the central role of the neuron may be detrimental for progress in neuroscience. The specialization of research into the subcomponents of the neuron has led to an excessive segmentation of fields, overlooking the unity of the neuron: "This failure to consider the neuron as a whole is not merely of historical significance, but of potential importance to the development and direction of clinically relevant strategies".

The neurosemantics project of this book is grounded on the neural doctrine, mostly for pragmatic reasons. While neurocomputation built on Cajal's legacy is now mature enough to offer tools that are suitable for modeling high level cognitive functions, which will be the content of Chap. 4, nothing similar exists today, that is based on or takes different computational bases, such as dendrites, into consideration. However, we must keep in mind, that all the models presented in this book, are based on a paradigm that can change drastically, as the result of better accounts of neurosemantics founded on a mathematics of the brain, yet to be discovered.

## 2.2 Plasticity

We have just discussed how the mathematical description of the neuron electrotonus achieved half a century ago, has been a major step, yet insufficient to fully characterize the computational properties of a neuron. One reason is in the cooperative behavior of neurons, whose meaningful functions arise only in assemblies of a large number of cells. A further, more compelling reason, is that even a formal model exactly describing all the chemical and electrical interactions of a group of neurons, with all the necessary parameters for reproducing the ongoing electrical activity at a given time, would still reveal an incomplete picture. It would be missing the fundamental aspect of the continuous self-modification of that group of neurons, driven by the ongoing activity itself. It is the ability of assemblies of neurons to modify their structural connectivity based on their own neural activity, that constructs meaningful computational functions. In humans, most of the organization of the brain at birth is immature, particularly in the cortex, and it is through the continuous interplay between the experience of patterned electrical signals and consequent modifications, that cognitive as well as non cognitive functions mature. Collectively, this process goes under the term of "plasticity". It embraces a number of distinct phenomena, such as axon arborization, dendrite rewiring, and the strengthening or weakening of local connections between two neurons: the synapse.

While in clinical neurology a large interest in plasticity focuses on the massive rewiring of connectivity in reaction to brain injury (Møller 2006; Fuchs and Flügge 2014), for the study of neurosemantics, synaptic plasticity is by far the most important. It is also the form of plasticity best known today, and the only one whose way of working has been schematized in computational terms. Therefore, this section will deal with this crucial boundary between one neuron and the next, the synapse. Its eminent role in intelligent behavior has roused curiosity on its origin and evolution, and motivated explorations even more challenging than those on ion channels and the neuron (see Sect. 2.1). According to Ryan and Grant (2009) the *ursynapse*, a kind of ancestor of all synapses, appeared in choanoflagellates about one billion years ago, evolving in *protosynapse* similar to the extant ones in cnidarian, around 700 millions years ago.

### 2.2.1 *The Elusive Passage*

As reported in Sect. 2.1.1 the discovery of the synapse is attributed to Sherrington (1906), who was well aware of its importance, but had no way, at the time, of exploring such a tiny space in order to understand its mechanisms. It is the crucial passage where the traveling action potential in an axon ends, and the dendrite of another neuron begins. For a long time, what exactly happens in this infinitesimal gap remained a mystery. In the years following Sherrington's discovery, two different hypotheses on the nature of the communication passing through the synapse gave rise to a vigorous scientific debate. On one side Henry Dale and Otto Loewi conjectured that the communication across synapses was chemical in nature. The best evidence was obtained in their experiments on the vagus nerve that controls heart rate, which led to the recognition that the substance released by the axon in the synapse is acetylcholine (Dale 1935). On the opposite side, John Eccles hypothesized that the current produced by an action potential in the axon crosses the synapse and directly enters the dendrite. His skepticism concerning chemical transmission derived from his conviction that signaling between neurons was too fast to be chemical in nature. This controversy become humorously referred to as the "soup versus spark" dustup, where clearly "soupers" sided with the chemical hypothesis, and "sparkers" with the electical one. Despite the increasing evidence in favor of the chemical theory of synaptic transmission, Eccles (1945) continued to defend his ideas and to experiment with electrical transmission even further. His theory was conclusively falsified by new evidence, in particular Katz (1959) went on to show the existence of ion channels that, unlike the voltage-gated channels of the axon, change their permeability in response to specific chemical transmitters only.

In Fig. 2.4 a simple scheme of the basic working of the synapse is given. The upper part, the presynapse, is the termination of an axon, and the lower part, the postsynapse, is a dendritic spine, a typical protuberance of the dendrite just in front of the presynapse. The space in between is the synaptic cleft, as thin as 15–25 nm (Peters et al. 1991). In the presynaptic termination there are small, spherical membrane-bounded organelles called *vesicles*, filled with neurotransmitters, the chemical messengers. The communication process is initiated when an action potential reaches the terminal of the presynaptic neuron. The change in membrane potential caused by the arrival of the action potential leads to the opening of

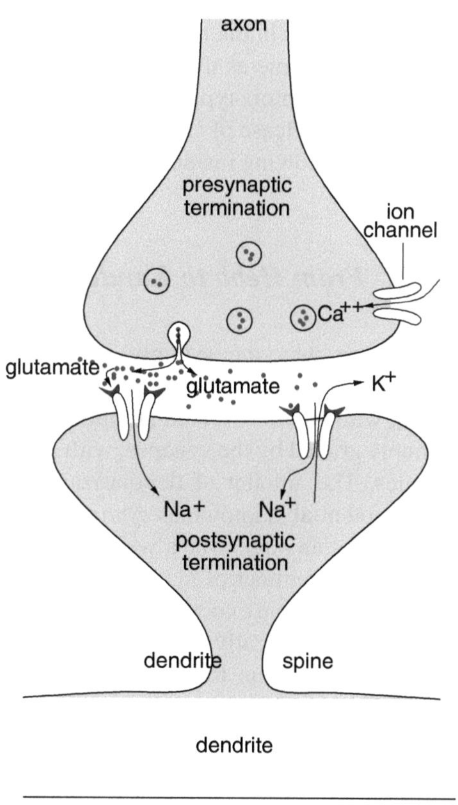


**Fig. 2.4** Scheme of an excitatory synapse. The termination of the axon is populated with neurotransmitters, packaged into *vesicles*, small membrane-enclosed organelles. When an action potential enters a presynaptic terminal, it causes calcium channels to open, letting $Ca^{++}$ ions flow into the cell. The local high density of $Ca^{++}$ induces the *exocytosis* of vesicles (the fusion of their membrane with the plasma membrane). The neurotransmitters diffuse across the synaptic cleft, and bind to receptors on the surface of the postsynaptic cell, triggering the opening of the non-selective ion channels

voltage-gated calcium channels, causing a rapid influx of $Ca^{++}$ ions. The elevation of $Ca^{++}$ concentration, in turn, favors the fusion of synaptic vesicles with the plasma membrane of the axon, a process called *exocytosis*. Since the number of neurotransmitters in each vesicle is almost constant, the synaptic transmission is *quantal*, graduated in discrete amplitude with the minimum step corresponding to a single vesicle (Katz 1971). This process lasts a couple of milliseconds.

On the opposite side of the synapse, the released neurotransmitters bind to the receptors placed on the surface of the dendrite spine. The binding induces conformational changes in the receptors that open the channel, normally impermeable, permitting ions to flow. The channel is not selective, therefore, both $K^+$ and $Na^+$ are allowed to flow, but the influx of $Na^+$ dominates, since at rest state there is little driving force on $K^+$. The net effect is a rapid local depolarization of the dendrite.

There are more than 100 types of different known neurotransmitters, even if the most abundant are of only a few types. We have already mentioned acetylcholine, the first discovered neurotransmitter, it is the chief chemical messenger for peripheral axons projecting in muscle fibers, and less important in the brain, where glutamate abounds. A first classification of neurotransmitters is based on their effect on the postsynaptic neuron, that could be excitatory, as in the scheme of Fig. 2.4, or inhibitory, in that the release of the neurotransmitter in the synapse decreases the likelihood of a postsynaptic action potential occurring. The most common inhibitory neurotransmitter in the brain is GABA (gamma-aminobutyric acid), its mechanism is exactly the same as that of an excitatory neurotransmitter. The difference is that the GABA receptors typically open channels that are selectively permeable to $Cl^-$, therefore, the release of GABA in the synaptic cleft has the final effect of negatively charged $Cl^-$ flowing inside the dendrite, producing hyperpolarization.

### 2.2.2 *From Hebb to Kandel*

The reason why the synapse is the chief location of plasticity, is due to its modulatory effect on the action potential, which by itself is a rather stereotypical signal, with an almost fixed amplitude. The binary value of the action potential can be finely graded by the synapse, with a quantum given by the size of the presynaptic vesicles. The amount of depolarization induced in the dendrite by a presynaptic action potential is known as *synaptic efficiency*, sometimes also known as *synaptic strength*, it can vary over a wide range. The flexibility is given by several factors, basically the number of synaptic vesicles ready for release, and the likelihood of a vesicle undergoing exocytosis at the arrival of an action potential. All these factors can change drastically depending on the previous history of the neurons at both synaptic sides. This is synaptic plasticity, and it is the key factor that provides the brain with the astonishing ability to forge the behavior of the organism in response to the environment, and to improve its performance over time, thanks to experience. From a mathematical point of view, ideally, sectioning a network of neurons in the brain, isolating its input and output connections, would produce a huge number of possible transfer functions between inputs and outputs, thanks to the degrees of freedom of the synaptic strengths between all internal neural connections. Among all the possible functions, the system tends to evolve towards one only, by automatically adapting the synaptic strengths, in reaction to the experienced patterns of activity.

How the past events of neurons exactly modify the parameters of the synapse, or mathematically, how the history of input patterns functionally relate to current neural function, is still largely unknown, and extremely hard to investigate. Long before the possibility of empirically investigating synaptic plasticity was made available, a brilliant intuition was provided, known today as "Hebb's law", and is defined in the following statement:

> When an axon of cell A is near enough to excite cell B and repeatedly or persistently takes part in firing it, some growth process or metabolic change takes place in one or both cells such that A's efficiency, as one of the cells firing B, is increased.
>
> (Hebb 1949, p. 62)

It is the coincidence in the timing of the activation of both presynaptic and postsynaptic neurons that produces an increase in the synaptic efficiency. The rule predicted by Hebb is of paramount relevance for explaining the capacity of the brain to represent the world, as will be discussed in Sect. 3.2.

Just as the squid has been uniquely useful in revealing the functioning of the axon electrotonus, the *Aplysia*, a small snail, was the animal that first allowed an initial understanding of synaptic plasticity. In this animal neurons are so large and uniquely identifiable, that it was feasible, even if extremely laborious, to track changes in synaptic communication. One behavior in particular, the gill-withdrawal reflex, is so simple that it engages no more than 24 neurons. The retraction of the gill, the breathing organ of the *Aplysia*, can be induced by touching the syphon, which is a spout that expels waste from the body. This reaction is flexible, and in particular, can be modified in two ways: habituation and sensitization. The former is the weakening of the reflex after repeated light touches, that are recognized as harmless. On the contrary, sensitization arises when the same light touch is associated with a noxious shock, after learning, a simple light touch produces a strong gill-withdrawal reflex. Eric Kandel led a group that obtained the first insights into the neural and molecular mechanisms of plasticity in the *Aplysia*, which won them the Nobel Prize in 2000. They found a direct correlation of habituation and sensitization with, respectively, the weakening and the strengthening of the synaptic connections in the gill-withdrawal circuit, the duration of this short-term memory is dependent on the length of time a synapse is weakened or strengthened (Kupfermann et al. 1972; Carew et al. 1972; Pinsker et al. 1973). Moreover, the two different kinds of learning differ in the ways the synapses are modified. During habituation the same neurons involved in the main control loop are active, and the change is homosynaptic, while in the case of sensitization, the shock is perceived by a neuron that is different from those of the gill-withdrawal reflex, and in this case, changes are heterosynaptic. Note that neither case corresponds directly to Hebb's law.

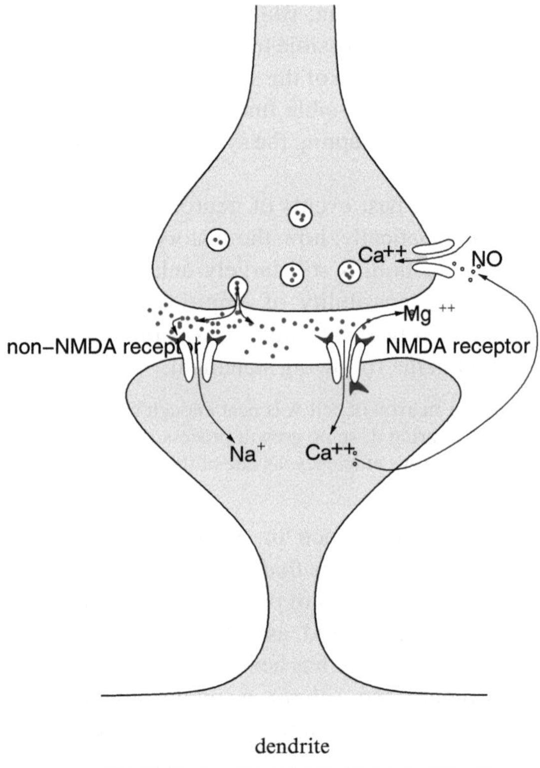


**Fig. 2.5** Scheme of a synapse with an NMDA receptor

The most studied physiological model of plasticity resembling what Hebb predicted is long-term potentiation (LTP) (Bliss and Lømo 1973; Artola and Singer 1987; Bliss and Collingridge 1993; Bear and Kirkwood 1993). The best known molecular mechanism for LTP relies upon NMDA receptors. NMDA stands for N-methyl-D-aspartate, an *agonist* that inactivates a specific subfamily of glutamate receptors. A convenient way to classify the variety of brain glutamate receptors is by naming their pharmacological agonists (Hollmann and Heinemann 1994), for the purpose of studying plasticity a convenient classification is just through NMDA and non-NMDA receptors. The unique properties of NMDA receptors combine two features: the receptor must bind glutamate in order to open the channel, but it is also voltage-gated, like axon $Na^{+}$ ion channels. As shown in Fig. 2.5, the dependence on the postsynaptic depolarization is due to a $Mg^{++}$ block site. When the internal potential reaches a threshold, the $Mg^{++}$ is released, and the channel can open, if there is glutamate to bind in the synaptic cleft. This double constraint matches Hebb's condition: the coincidental firing of the presynaptic and the synaptic neurons. When the NMDA receptor opens, it allows not only $Na^{+}$ and $K^{+}$ to flow, but also $Ca^{++}$ ions. The increase of intracellular $Ca^{++}$, as seen in the axon terminal, activates a variety of chemical processes, whose final effect is a strengthening of synaptic efficiency. The synaptic changes may involve the presynaptic termination

**Table 2.1** A gallery of synaptic plasticities. The presynaptic termination is at the top, the postsynaptic in the bottom, both are shown in light gray when active, in dark otherwise. The "+" or "−" indicate increase or decrease in synaptic strength

| Hebbian | Anti-Hebbian | Homosynaptic | Heterosynaptic |
|---|---|---|---|
| 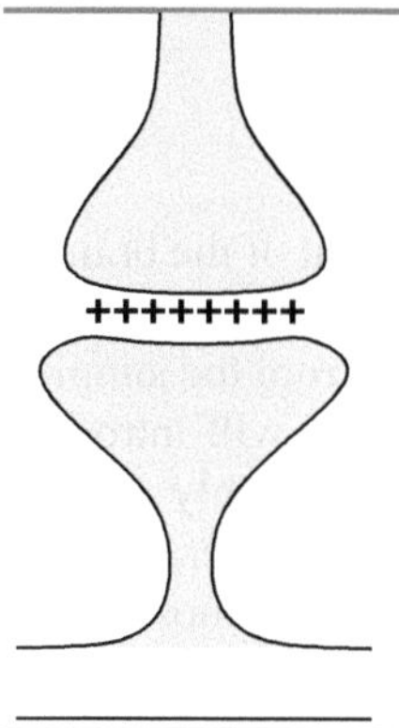 | 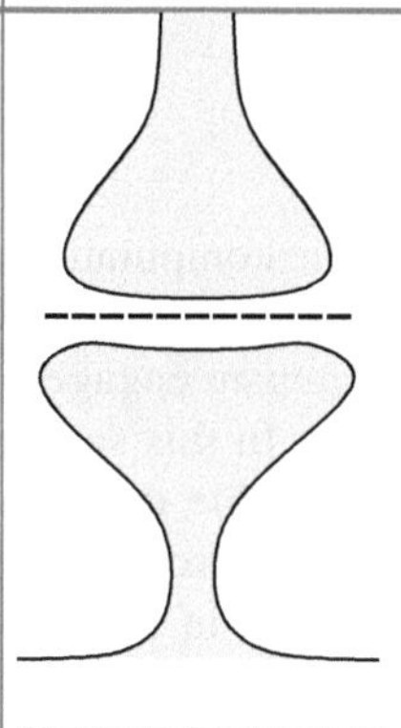 | 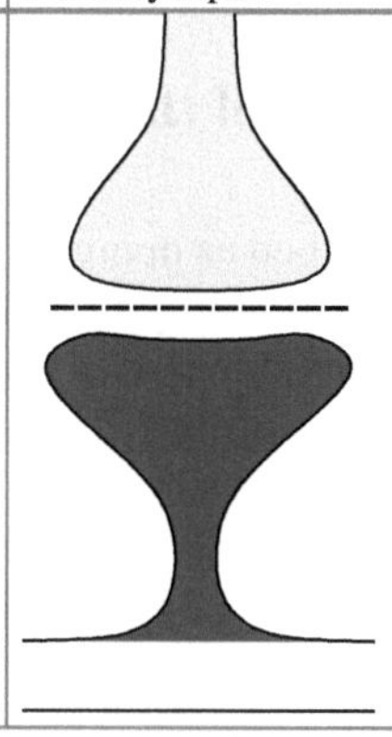 | 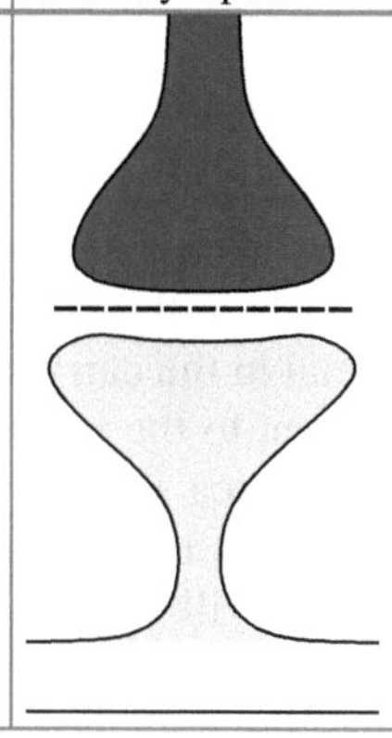 |

as well, thanks to a *retrograde messenger*, a chemical compound traveling from the dendrite to the axon. NO (Nitric Oxide) has been one of the first suggested as a possible candidate, being freely diffusible and generated as a result of high $Ca^{++}$ concentration. Because of its precious role, it won the "Molecule of the Year" Award in 1992 (Koshland 1992), but its exact behavior turns out to be quite complex, and still remains controversial (Susswein et al. 2004).

In summary, Hebb was right, but he envisaged just one among the many ways of changing synaptic strengths in the brain. Long-term depression (LTD) is the converse process to LTP and results in a long lasting decrease in synaptic efficacy. The main cause of synaptic weakening is deprivation of presynaptic or postsynaptic activities (Ito 1989; Zhuo and Hawkins 1995). LTD has also been observed in synapses with NMDA receptors (Crozier et al. 2007). LTP and LTD are far from encompassing the full range of plasticity at the synaptic level, a synopsis of synaptic modification phenomena is given in Table 2.1. The plot on the left is a synapse that exactly follows Hebb's law. Crucially, the opposite phenomena has been observed as well: synaptic weakening when the two neurons are both active: anti-Hebbian learning, shown in the next plot from left to right. The choice between the two behaviors seems to depend on the temporal order of the two firings: if the presynaptic action potential occurs before postsynaptic activation, within a few milliseconds, the Hebbian rule applies, if the postsynaptic neuron fires before the arrival of the action potential in the presynaptic axon, the anti-Hebbian rule takes over. This overall behavior is known as *spike-timing-dependent plasticity*, as argued by Markram et al. (2011), it is prone to interesting philosophical speculations (see also Sect. 3.2.3). It may appear as the brain correlate of our need to explain facts of the world in terms of causality. The synapse is induced to reinforce a representation, whenever a signal is a prediction of some other signal, otherwise it is taken as a false association to be discarded. The other two forms of synaptic modification shown in Table 2.1, homosynaptic and heterosynaptic, do not require the simultaneous activation of two cells in the synapse, and are those that were discovered by Eric Kandel, described above.

## 2.3 The Organization of the Cortex

The neuron has been described as being the basic computational unit of the brain. It is certainly at the core of the ability to code for meaning, but first and foremost, it is crucial in the carrying out of any task an organism engages in, from the simplest movement to the perception of its environment. In this section, we will introduce a special area and a way in which neurons assemble there, that is highly specific and efficient in the construction of meaning: the cerebral cortex. In fact, most of this book will deal with processes that take place in the cortex, and most of the computations proposed for the modeling of semantic processes will directly refer to those done by the cortex. This section will provide a preliminary introduction to what the cortex is and will discuss the kinds of cells and connections that compose it.

An early step in the history of brain evolution, long before the cortex appeared, was the clustering of a growing number of interneurons in the anterior part of the body, which lead to the formation of the brain. This is thought to have happened about 560 millions years ago, probably in the freshwater flatworm of the genus *Planaria* (Nakazawa et al. 2003). In the long and varied diversification of the brain throughout evolution, a major turning point occurred about 200 millions years ago, with the formation of a uniform superficial fold, composed of six layers (Striedter 2003). It was populated by a newly shaped neuron with a pyramidal form, suggestively named *psychic cell* by Ramón y Cajal (1906), who had already grasped their key role in constructing the mind.

### *2.3.1 The Origins of the Cortex*

It is well agreed upon that the mammalian neocortex is the site of processes enabling higher cognition, from consciousness to symbolic reasoning, and, especially relevant here, linguistic meaning (Miller et al. 2002; Farhat 2007; Fuster 2008; Nieder 2009; Noack 2012). There is much less consensus, however, on the reason why the particular way neurons are combined in the cortex makes such a difference with respect to the rest of the brain. Possible explanations will be reviewed in Sect. 3.3. Edinger (1904) was one of the first to rank mammals as the most intelligent animals, in virtue of the brand new layered brain equipment introduced by nature. This intellectual superiority remained almost undiscussed for a century (Romer 1967), clearly meeting with a certain amount of anthropocentric self-satisfaction. The rise of comparative cognition has weakened this certainty, with the discovery of abilities previously thought to be exclusive to mammals. Flexibility, the ability to learn new strategies as well as problem solve, have been reported in a wider set of animals and not restricted to mammals (Pearce 2008), and would include reptiles (Davidson 1966) and birds (Pepperberg 1999). Although it is not easy to precisely define in a sentence or two what it is exactly that distinguishes mammal intelligence from that of other species, the cortex is certainly considered to be the crowning achievement of brain evolution, and the quest for an understanding of its computational properties and its origins, are among the most prominent and yet unresolved issues in neurobiology.

Broadly speaking the cortex can be seen in continuation with the *pallium*, which by definition is the external folding of the brain in all vertebrates. In sauropods as well, it is the site of higher mental processes (Medina and Abellán 2009). However, the special six-layered structure is missing in all other classes except mammals, with no structure present at all in amphibians and a simple three-layered organization in the *pallium* of reptiles. More precisely, only a part of the entire cortex presents the new stratified feature, for this reason it is also called *neocortex*, or *isocortex* for its being highly uniform throughout its extension. There are two remaining parts that are similar to the old reptilian three-layered *pallium*: the hippocampus, involved in spatial localization and long term memory, and the piriform cortex, which processes olfactory signals. As can be seen in Fig. 2.6, the proportion of neocortex over the entire cortex varies among mammalian species, and becomes predominant in humans.

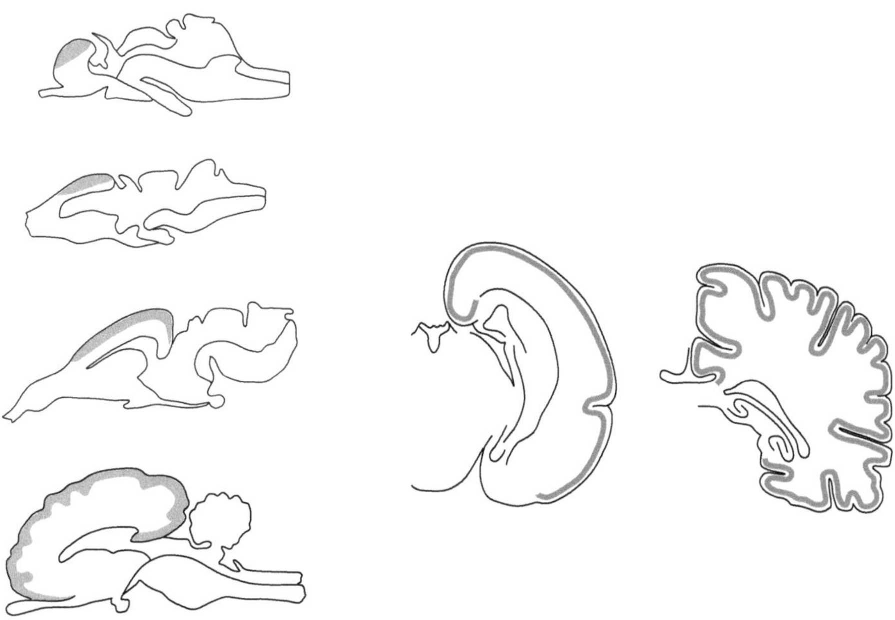

**Fig. 2.6** Extension and structure of the *pallium* in different animal species. In the *left column* a general comparison between the principle classes of vertebrates, starting from *the top*: fish, amphibian, reptile, mammal. In the *central column* and in the *right column*, detail of the coronal section, of the marsupial *Opossum* and of a human (Adapted from Fuster 2008)

The kind of brain reorganization that mutated the *pallium* into the neocortex is still unclear and controversial. There is an area in the reptilian brain, the DVR (*Dorsal Ventricular Ridge*), which enjoys the most favor as a possible precursor, due to its having five types of neurons resembling those populating the different layers of the neocortex (Butler 1994; Karten 1997). Others, like Ulinski (1990) and Aboitiz (1999), contend that DVR is more likely homologous to internal areas of mammalian brains, like the amigdala (Bruce and Neary 1995), while the neocortex is a derivation from the dorsal *pallium*. The lamination of the cortex, in this latter hypothesis, is explained by slower development, with some late-born phylogenetically new neurons, which participated in local processing circuits, as opposed to older neurons. The gradual migration of neurons from their proliferative zones towards the cortex is uniquely characterized by the *inside-out* neurogenetic gradient, in that each new generation bypasses the previous one in the cortex. In the expansion of the neocortex, the effect of mutations in some genes involved in the regulation of cell migration stabilized the distinct development of layers.

Once the neocortex structure was established in mammals, further evolution caused its expansion along the brain's surface. A macaque has a ratio of 100:1 of cortical surface normalized to the whole brain, compared to the mouse, humans have a ratio of 1000:1. An outstanding peculiarity of the enlargement of the cortex is that while it occured mainly through the expansion of the surface area, only a modest increase took place in its thickness. Apparently an increase in depth would not offer any substantial additional computational power. The need of enlarging the cortex by its superficial area only has led to its folding, culminating in the large and deep convolutions of the human cortex.

The striking parallel between the proportions of cortex and the whole brain during human development and mammalian evolution, that supports the so-called "evo-devo" concept (Maienschein 2007), has provided the opportunity of shedding light on how the cortex evolved in size, by investigating the mechanisms of corticogenesis. One of the best current explanations of why the cortex expanded along the surface of the brain without a comparable increase in its thickness is the radial unit hypothesis (Rakic 2009). In the proliferative embryonic zones, such as the ventricular zone, neural stem cells grow by symmetrical division before the onset of neurogenesis, This means that each stem cell divides in two, with each potentially being a founder cell that gives rise to a radial cortical column, but before being transformed into a founder cell it can divide again in two. In this way an exponential number of potential cortical columns is produced. After the onset of neurogenesis each founder cell will divide asymmetrically, inducing a linear growth inside a column. By genetic manipulation an increase in the precursor population has been induced in the mouse, which results in an increased number of radial columns and consequently a convoluted cortex (Chenn and Walsh 2002) (Fig. 2.6).

### 2.3.2 *Circuit Structure*

Despite its huge extent, and the range of different functions carried out by its areas, the cortex is amazingly uniform. The invariance of its basic microstructure was already known and described by its first historical investigators (Ramón y Cajal 1906; Brodmann 1909). In the cortex the neural density is extremely stable. Rockel et al. (1980) counted about 110 neurons in sections of 30 $\mu$m diameter of cortex, either in motor, somatosensorial, frontal, parietal, or temporal areas, across animals such as mice, cats, monkeys, and humans. The only exception is always to be found in the primary visual cortex, with a count of about 270 neurons. Their observations have been the subject of fierce debate for over 30 years, with doubts raised concerning whether their experimental methods were technically flawed (Rakic 2008), but recently Carlo and Stevens (2013) carefully replicated their experiments confirming previous results. Moreover, Karbowski (2014) found a number of additional parameters that are remarkably constant across species and across regions, like that of adult synaptic density, with a mean of $5 \times 10^{11}$ cm$^{-3}$, and the ratio of excitatory to inhibitory synapses, around 5.6.

One of the most important and studied uniform aspects of the cortex is its layered organization, with the overlap of laminae composed by different types of cells, myelination, and pigmentation. The delineation of the six distinct layers were first noted by Berlin (1858), the details of the layers, with a blueprint that is still mostly valid today, were revealed by early neuroscientists such as Ramón y Cajal (1906), Brodmann (1909), Vogt and Vogt (1919), and von Economo and Koskinas (1925). The layers are listed in the following table, using Brodmann's Latin nomenclature and the one recommended by Vogt and Vogt. Layers identified by cytoarchitecture are usually marked in Roman numerals, while those with distinct myelin in Arabic numerals.

| Cytoarchitectonics layers | | | Myeloarchitectonics layers | |
|---|---|---|---|---|
| I | *Lamina zonalis* | Molecular layer | Zonal layer | 1 |
| II | *Lamina granularis externa* | Corpuscular layer | Dysfibrous layer | 2 |
| III | *Lamina pyramidalis* | Pyramidal layer | Suprastriate layer | 3 |
| IV | *Lamina granularis interna* | Granular layer | External stria | 4 |
| V | *Lamina ganglionaris* | Ganglionic layer | Internal stria | 5 |
| VI | *Lamina multiformis* | Multiform layer | Substriate layer | 6 |

The cytoarchitectonic nomenclature typically derives from a layer being prevalently populated by one or a number of the many types of cells, which will be described in some detail below. The first layer actually lacks a specific type, due to its almost total lack of cells, in fact, they are very few and scattered. It is filled by terminal ramifications of axons and dendrites of neurons in other layers. The *corpuscular* qualification of the second layer comes from neurons there being quite small and tightly packed. The pyramidal layer, making up almost one third of the total thickness of the cortex, is the site of the largest well-formed pyramidal cells. The fourth layer is similar to the second, in that cells are small and densely packed, but it is variable in size depending on the cortical region. This layer is the main target of thalamocortical projections, as well as intra-hemispheric corticocortical connections, therefore, it is well developed in sensorial areas and reduced in regions with scarce thalamic inputs, like motor areas. The ganglionic layer is also region dependent, for the opposite reason: it is mainly populated by pyramidal cells projecting into the basal ganglia or directly to the corticospinal tract, and is therefore highly developed in all motor areas. The multiform layer is prevalently composed of densely collected spindle-shaped cells, with their long axes arranged perpendicularly to the cortical surface. The different extent of layers IV and V has been used by von Economo and Koskinas (1925) for a broad classification of the cortex into *granular*, typical of sensorial areas, and *agranular*, such as the motor areas. Collectively, the two types form the *heterotypical* cortex, the remaining area is the *homotypical* cortex, which is not primarily engaged in sensorial processing or in motor output. It is the core of higher level cognition, abundant in the frontal lobe.

The principal neuron in the cortex is the pyramidal cell, Cajal's *psychic cell*, accounting for 70 % of the overall neural population in the cortex, prevalent in layers II, III, and V. Their body is conical with the vertex directed toward the surface, carrying the apical dendrite, a channel with rich terminal tuft into the superficial layers and additional synaptic domains placed depending on the level at which the soma is situated. A second set of dendrites radiates its bouquet from the base of the soma. All dendrites are covered with spines, bumps of $0.1\ \mu m^3$ height increasing the efficiency of synaptic transmission.

Most of the axons of pyramidal cells leave the cortex, and represent the main cortical output, with targets depending on the layer of the cells. Cells in layers II and III project to other cortical areas, either ipsilaterally or via the corpus callosum, in the other hemisphere. As previously mentioned, motor output to basal ganglia and the spinal cord come from pyramidal cells in layer V. The output of pyramidal neurons in layer VI is mainly directed to thalamic nuclei (Fig. 2.7).

A prominent feature of pyramidal neurons is the conspicuous number of intracortical collateral branches, constituting by far the largest source of excitatory signals in the cortex. Lateral connections are well myelinated and spread to distances up to several millimeters, often in periodically organized groups of synapses (Gilbert et al. 1990; Hou et al. 2003). There is strong evidence of the relevant computational role of lateral connections, discussed in detail in Sect. 3.3.2.

There are other excitatory cortical neurons, that can be considered, at least in part, as modified pyramidal cells. The most important is the *spiny stellate*, which is concentrated in layer IV, producing its "granular" aspect, and is abundant in the visual cortex. The input to spiny stellate neurons is fed primarily by thalamic fibers, with secondary contributions from pyramidal cells of the deeper layers, or other spiny stellate. Their main computational contribution seems strongly related to the propagation of thalamic signals. The *bipolar* cells are less frequent, and have

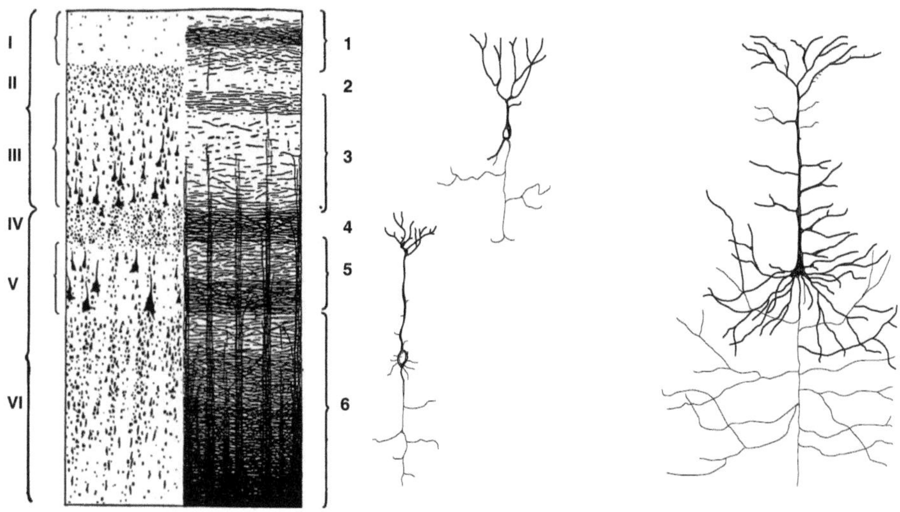


**Fig. 2.7** In the figure *on the left*, an original image of Brodmann (1909), the cellular and myelinic structure of the cortex is illustrated. The figure *on the right* shows a series of original drawings made by Cajal Ramón y Cajal (1906), comparing philogenesis and ontogenesis: from the left, a human neuroblast with emerging basal dendrites and axon collateral branches, a mature lizard cell, a mature human cell

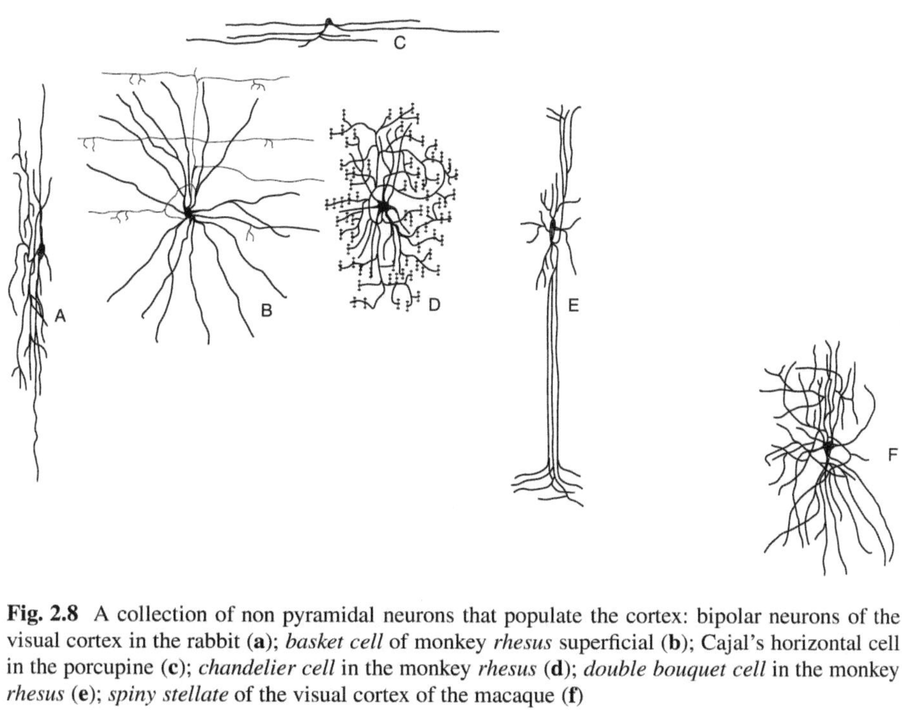


**Fig. 2.8** A collection of non pyramidal neurons that populate the cortex: bipolar neurons of the visual cortex in the rabbit (**a**); *basket cell* of monkey *rhesus* superficial (**b**); Cajal's horizontal cell in the porcupine (**c**); *chandelier cell* in the monkey *rhesus* (**d**); *double bouquet cell* in the monkey *rhesus* (**e**); *spiny stellate* of the visual cortex of the macaque (**f**)

a typically narrow and radially elongated shape, which is almost symmetrical to the soma. Their main function, therefore, seems to be the radial communication of signals.

Inhibitory interneurons are sharply different from pyramidal neurons, GABA is their main neurotransmitter, and they almost entirely lack dendritic spines. The most common is the *basket cell*, concentrated in layers III and V, with poorly ramifying dendrites, and axons with very short radial lengths, giving rise to very long lateral connections. Their targets are pyramidal cells and spiny stellate as well. The *chandelier cell*, common in layer II, owe their name to the axonal plexus carrying a large number of radial axonal swellings, resembling candles. They seem to be the most influential source of inhibition for pyramidal cells. The *double bouquet cell* also earned a suggestive name, from Cajal, due to the shape of its axons, made up by double, thin and long radial branches, terminating in two distant tufts. The last cell to be mentioned here is the *horizontal cell of Cajal*, the only cell which includes its baptizer in its name. It is the only cell with a displacement that is entirely parallel to the cortical surface. Confined exclusively to layer I, it is another component of the strategic lateral interaction in the cortex.

Figure 2.9 shows a diagram of the typical circuital connections of the cortex, that involve the neurons described as well as other minor ones, giving place to a complex local cortical circuitry that is periodically replicated parallel to the surface, spaced between 300 and 800 $\mu$m. This gives rise to the so called *columnar* organization of the cortex, that will be described in depth in Sect. 3.3.1.

The composition and basic structure of the cortex, here briefly sketched out, is very likely the best computational organization nature has reached so far. It is the most plausible explanation of the universality of its design across brain regions and

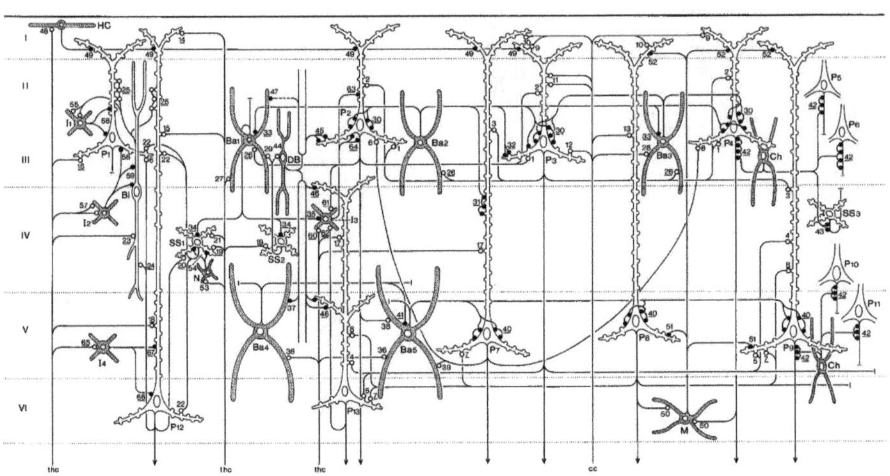


**Fig. 2.9** Diagram of a typical circuit in the cortex, reconstructed from the study of experimental cortical slices (From Nieuwenhuys 1994)

species. There are clearly local variations in certain aspects of the cortex, seen for example in the differences between granular perceptual areas and agranular motor areas, and the taxonomy has been subtly refined in current neuroanatomy (Braak 1980). But all these variations are marginal with respect to the basic main layered architecture, made up of the populations of neurons here described. This universal uniformity has motivated the search for the fundamental neuronal circuit that has proven to be so successful. It will be the topic of the sections that follow.

## References


Aboitiz, F. (1999). Comparative development of the mammalian isocortex and the reptilian dorsal ventricular ridge. evolutionary considerations. *Cerebral Cortex, 9*, 783–791.

Artola, A., & Singer, W. (1987). Long term potentiation and NMDA receptors in rat visual cortex. *Nature, 330*, 649–652.

Azmitia, E. C., DeFelipe, J., Jones, E. G., Rakic, P., & Ribak, C. E. (Eds.). (2002). Changing Views of Cajal's Neuron. *Progress in Brain Research* (Vol. 136). Amsterdam: Elsevier.

Bain, A. (1873). *Mind and body. The theories of their relation*. London: Henry King.

Baluška, F., Volkmann, D., & Menzel, D. (2005). Plant synapses: Actin-based domains for cell-to-cell communication. *TRENDS in Plant Science, 10*, 106–111.

Bear, M., & Kirkwood, A. (1993). Neocortical long term potentiation. *Current Opinion in Neurobiology, 3*, 197–202.

Berlin, R. (1858). Beitrag zur structurlehre der grosshirnwindungen. PhD thesis, Medicinischen Fakultät zu Erlangen

Bliss, T., & Collingridge, G. (1993). A synaptic model of memory: Long-term potentiation in the hippocampus. *Nature, 361*, 31–39.

Bliss, T., & Lømo, T. (1973). Long-lasting potentiation of synaptic transmission in the dentate area of the anaesthetized rabbit following stimulation of the perforant path. *Journal of Physiology, 232*, 331–356.

Bogen, J. (2005). Regularities and causality; Generalizations and causal explanations. *Studies in History and Philosophy of Science Part C, 36*, 397–420.

Braak, H. (1980). *Architectonics of the human telencephalic cortex*. Berlin: Springer.

Brenner, E. D., Stahlberg, R., Mancuso, S., Vivanco, J., Baluška, F., & Volkenburgh, E. V. (2006). Plant neurobiology: An integrated view of plant signaling. *TRENDS in Plant Science, 11*, 413–419.

Brodmann, K. (1909). *Vergleichende Lokalisationslehre der Grosshirnrinde*. Leipzig: Barth.

Bruce, L. L., & Neary, T. J. (1995). The limbic system of tetrapods: A comparative analysis of cortical and amygdalar populations. *Behavioral and Brain Science, 46*, 224–234.

Bullock, T. H., Bennett, M. V. L., Josephson, D. J. R., Marder, E., & Fields, R. D. (2005). The neuron doctrine, redux. *Science, 310*, 791–793.

Bush, P. C., & Douglas, R. J. (1991). Synchronization of bursting action potential discharge in a model network of neocortical neurons. *Neural Computation, 3*, 19–30.

Butler, A. B. (1994). The evolution of the dorsal pallium in the telencephalon of amniotes: Cladistic analysis and a new hypothesis. *Brain Research Reviews, 19*, 66–101.

Cao, R. (2011). A teleosemantic approach to information in the brain. *Biology and Philosophy, 27*, 49–71.

Cao, R. (2014). Signaling in the brain: In search of functional units. *Philosophy of Science, 81*, 891–901.

Carew, T. J., Pinsker, H. M., & Kandel, E. R. (1972). Long-term habituation of a defensive withdrawal reflex in Aplysia. *Science, 175*, 451–454.

Carlo, C. N., & Stevens, C. F. (2013). Structural uniformity of neocortex, revisited. *Proceedings of the Natural Academy of Science USA, 110*, 719–725.
Chenn, A., & Walsh, C. A. (2002). Regulation of cerebral cortical size by control of cell cycle exit in neural precursors. *Science, 297*, 365–369.
Connors, B. W., & Long, M. A. (2004). Electrical synapses in the mammalian brain. *Annual Review of Neuroscience, 27*, 393–418.
Craver, C. F. (2007). *Explaining the brain: Mechanisms and the mosaic unity of neuroscience*. Oxford: Oxford University Press.
Crozier, R. A., Wang, Y., Liu, C. H., & Bear, M. F. (2007). Deprivation-induced synaptic depression by distinct mechanisms in different layers of mouse visual cortex. *Proceedings of the Natural Academy of Science USA, 104*, 1383–1388.
Dale, H. H. (1935). Pharmacology and nerve endings. *Proceedings of the Royal Society of Medicine, 28*, 319–332.
Davidson, R. S. (1966). Operant stimulus control applied to maze behavior: Heat escape conditioning and discrimination reversal in alligator mississippiensis. *Journal of the Experimental Analysis of Behavior, 9*, 671–676.
Deiters, O. F. K. (1865). *Untersuchungen über Gehirn und Rückenmark des Menschen und der Säugethiere*. Braunschweig (DE): Vieweg.
Dretske, F. I. (1981). *Knowledge and the flow of information*. Cambridge: MIT.
du Bois-Reymond, E. (1849). *Untersuchungen über Thierische Elektricität*. Berlin: G. Reimer.
Eccles, J. C. (1945). An electrical hypothesis of synaptic and neuromuscular transmission. *Nature, 156*, 680–682.
Edinger, L. (1904). *Vorlesungen über den Bau der nervösen Zentralorgane des Menschen und der Tiere*. Leipzig (DE): Vogel.
Farhat, N. H. (2007). Corticonic models of brain mechanisms underlying cognition and intelligence. *Physics of Life Reviews, 4*, 223–252.
Fields, D., & Stevens-Graham, B. (2002). New insights into neuron-glia communication. *Science, 298*, 556–562.
Freud, S. (1885). Eine neue Methode zum Studium des Faserverlaufs in Centralnervensystem. *Centralbl Med Wissensch, 22*, 461–512.
Freud, S. (1895). *Project for a scientific psychology*. London: The Hogarth Press.
Fuchs, E., & Flügge, G. (2014). Adult neuroplasticity: More than 40 years of research. *Neural Plasticity, 2014*, ID541870.
Fukuda, T. (2009). Network architecture of gap junction-coupled neuronal linkage in the striatum. *Journal of Neuroscience, 29*, 1235–1243.
Fuster, J. M. (2008). *The prefrontal cortex* (4th ed.). New York: Academic.
Garz6n, P. C., & Keijzer, F. (2011). Plants: Adaptive behavior, root-brains, and minimal cognition. *Adaptive Behavior, 19*, 155–171.
Gerlach, J. (1871). Von dem Rückenmark. In S. Stricker (Ed.), Handbuch der Lehre von den Geweben des Menschen und der Thiere, W. Engelmann, Leipzig (DE).
Gerstner, W. (1999). Spiking neurons. In W. Maass & C. M. Bishop (Eds.). *Pulsed neural networks*. Cambridge: MIT.
Gilbert, C. D., Hirsch, J. A., & Wiesel, T. N. (1990). Lateral interactions in visual cortex. In *Cold spring harbor symposia on quantitative biology* (pp. 663–677). New York: Cold Spring Harbor Laboratory Press.
Goldman, D. (1943). Potential, impedence, and rectification in membranes. *The Journal of General Physiology, 27*, 37–60.
Golowasch, J., Buchholz, F., Epstein, I. R., & Marder, E. (1992). Contribution of individual ionic currents to activity of a model stomatogastric ganglion neuron. *Journal of Neurophysiology, 67*, 341–349.
Hebb, D. O. (1949). *The organization of behavior*. New York: John Wiley.
Hille, B. (1992). *Ionic channels of excitable membranes*. Sunderland: Sinauer

Hodgkin, A. L., & Huxley, A. F. (1952). A quantitative description of ion currents and its applications to conduction and excitation in nerve membranes. *Journal of Physiology, 117*, 500–544.

Hollmann, M., & Heinemann, S. (1994). Cloned glutamate receptors. *Annual Review of Neuroscience, 17*, 31–108.

Hoorweg, J. L. (1889). Ueber die elektrischen Eigenschaften der Nerven. *Pflügers Arch ges Physiol, 71*, 128–157.

Hou, C., Pettet, M. W., Sampath, V., Candy, T. R., & Norcia, A. M. (2003). Development of the spatial organization and dynamics of lateral interactions in the human visual system. *Journal of Neuroscience, 23*, 8630–8640.

Ito, M. (1989). Long-term depression. *Annual Review of Neuroscience, 12*, 85–102.

Kaplan, D. M., & Craver, C. F. (2011). Towards a mechanistic philosophy of neuroscience. In S. French & J. Saatsi (Eds.), *Continuum companion to the philosophy of science* (pp. 268–292) London: Continuum Press.

Karbowski, J. (2014). Constancy and trade-offs in the neuroanatomical and metabolic design of the cerebral cortex. *Frontiers in Neural Circuits, 8*, 9.

Karten, H. J. (1997). Evolutionary developmental biology meets the brain: The origins of mammalian cortex. *Proceedings of the Natural Academy of Science USA, 94*, 2800–2804.

Katz, B. (1959). Mechanisms of synaptic transmission. *Reviews of Modern Physics, 31*, 524–536.

Katz, B. (1971). Quantal mechanism of neural transmitter release. *Science, 173*, 123–126.

Kelvin, W. T. (1855). On the theory of the electric telegraph. *Proceedings of the Royal Society of London, 7*, 382–399.

Kistler, W. M., Gerstner, W., & van Hemmen, L. (1997). Reduction of the Hodgkin-Huxley equations to single-variable threshold model. *Neural Computation, 9*, 1015–1045.

Kölliker, A. (1893). Über die feinere Anatomie und die physiologische Bedeutung des sympathischen Nervensystems. *Wiener klin Wochenschr, 7*, 773–776.

Koshland, D. E. (1992). The molecule of the year. *Science, s58*, 1861.

Kuhn, T. (1962). *The structure of scientific revolutions* (2nd ed., 1970). Chicago: Chicago University Press.

Kupfermann, I., Castellucci, V., Pinsker, H. M., & Kandel, E. R. (1972). Neuronal correlates of habituation and dishabituation of the gill-withdrawal reflex in Aplysia. *Science, 167*, 1743–1745.

Liebeskind, B. J., Hillisa, D. M., & Zakona, H. H. (2011). Evolution of sodium channels predates the origin of nervous systems in animals. *Proceedings of the Natural Academy of Science USA, 108*, 9154–9159.

London, M., & Häusser, M. (2005). Dendritic computation. *Annual Review of Neuroscience, 28*, 503–5032.

Maienschein, J. (2007). To Evo-Devo through cells, embryos, and morphogenesis. In M. D. Laubichler & J. Maienschein (Eds.), *From embryology to Evo-Devo* (pp. 109–121). Cambridge: MIT.

Markram, H., Gerstner, W., & Sjöström, P. J. (2011). A history of spike-timing-dependent plasticity. *Frontiers in Synaptic Neuroscience, 3*, 4.

McCulloch, W., & Pitts, W. (1943). A logical calculus of the ideas immanent in nervous activity. *Bulletin of Mathematical Biophysics, 5*, 115–133.

Medina, L., & Abellán, A. (2009). Development and evolution of the pallium. *Seminars in Cell & Developmental Biology, 20*, 698–711.

Miller, E. K., Freedman, D. J., & Wallis, J. D. (2002). The prefrontal cortex: Categories, concepts and cognition. *Philosophical Transactions: Biological Sciences, 357*, 1123–1136.

Møller, A. R. (Ed.). (2006). *Neural plasticity and disorders of the nervous system*. Cambridge: Cambridge University Press.

Moroz, L. L. (2009). On the independent origins of complex brains and neurons. *Brain, Behavior and Evolution, 74*, 177–190.

Nakazawa, M., Cebria, F., andKazuho Ikeo, K. M., Agata, K., & Gojobori, T. (2003). Search for the evolutionary origin of a brain: Planarian brain characterized by microarray. *Molecular Biology and Evolution, 20*, 784–791.
Neher, E., & Sakmann, B. (1976). Noise analysis of drug induced voltage clamp currents in denervated frog muscle fibers. *Journal of Physiology, 258*, 705–729.
Nieder, A. (2009). Prefrontal cortex and the evolution of symbolic reference. *Current Opinion in Neurobiology, 19*, 99–108.
Nieuwenhuys, R. (1994). The neocortex. *Anatomy and Embryology, 190*, 307–337.
Noack, R. A. (2012). Solving the "human problem": The frontal feedback model. *Consciousness and Cognition, 21*, 1043–1067.
Nordau, M. (1895). *Paradoxes psychologiques*. Paris: Alcan.
Pearce, J. M. (2008). *Animal learning and cognition: An introduction*. East Sussex: Psychology Press.
Pepperberg, I. M. (1999). *The alex studies: Cognitive and communicative abilities of grey parrots*. Cambridge: Harvard University Press.
Peters, A., Palay, S. L., & deF Webster, H. (1991). *Fine structure of the nervous system: Neurons and their supporting cells*. Oxford: Oxford University Press.
Pinsker, H. M., Hening, W. A., Carew, T. J., & Kandel, E. R. (1973). Long-term sensitization of a defensive withdrawal reflex in Aplysia. *Science, 182*, 1039–1042.
Rakic, P. (2008). Confusing cortical columns. *Proceedings of the Natural Academy of Science USA, 34*, 12099–12100.
Rakic, P. (2009). Evolution of the neocortex: A perspective from developmental biology. *Nature Reviews Neuroscience, 10*, 724–735.
Rall, W. (1967). Distinguishing theoretical synaptic potentials computed for different soma-dendritic distributions of synaptic input. *Journal of Neurophysiology, 30*, 1138–1168.
Ramón y Cajal, S. (1899). Textura del sistema nervioso del hombre y de los vertebrados (Vol I). Imprenta y Librería de Nicolás Moya, Madrid (English translation by P. Pasik and T. Pasik, 1997). Springer.
Ramón y Cajal, S. (1906). In J. DeFelipe & E. G. Jones (Eds.), *Cajal on the cerebral cortex: An annotated translation of the complete writings*. Oxford: Oxford University Press, 1988.
Rinzel, J. (1990). Electrical excitability of cells, theory and experiment: Review of the Hodgkin-Huxley foundation and an update. *Bulletin of Mathematical Biology, 52*, 5–23.
Rockel, A., Hiorns, R., & Powell, T. (1980). The basic uniformity in structure of the neocortex. *Brain, 103*, 221–244.
Romer, A. S. (1967). Major steps in vertebrate evolution. *Science, 158*, 1629–1637.
Ryan, T. J., & Grant, S. G. N. (2009). The origin and evolution of synapses. *Nature Reviews Neuroscience, 10*, 701–713.
Schwann, T. (1839). *Mikroskopische Untersuchungen über die Uebereinstimmung in der Struktur und dem Wachsthum der Thiere und Pflanzen*. Berlin: Sander.
Shannon, C. E., & Weaver, W. (1949). *The mathematical theory of communication*. Urbana: University of Illinois Press.
Sherrington, C. S. (1897). The mammalian spinal cord as an organ of reflex action. *Proceedings of the Royal Society of London, 61*, 220–221.
Sherrington, C. S. (1906). On the proprioceptive system, especially in its reflex aspect. *Brain, 29*, 467–482.
Sherrington, C. S. (1941). *The integrative action of the nervous system* (2nd ed., 8 Vols). Cambridge: Cambridge University Press.
Sidiropoulou, K., Pissadaki, E. K., & Poirazi, P. (2006). Inside the brain of a neuron. *EMBO reports, 7*, 886–892.
Stevens, C. (1994). What form should a cortical theory take? In C. Koch & J. Davis (Eds.), *Large-scale neuronal theories of the brain* (pp. 239–255). Cambridge: MIT.
Striedter, G. F. (2003). *Principles of brain evolution*. Sunderland: Sinauer Associated.
Sulloway, F. J. (1982). *Freud: Biologie der Seele: Jenseits der psycoanalytischen Legende*. Köln-Lövenich: Hohenheim Verlag.

Susswein, A. J., Katzoff, A., Miller, N., & Hurwitz, I. (2004). Nitric oxide and memory. *Neuroscientist, 10*, 153–162.

Traub, R. D., Wong, R. K. S., Miles, R., & Michelson, H. (1991). A model of CA3 hippocampal pyramidal neuron incorporating voltage-clamp data on intrinsic conductances. *Journal of Neurophysiology, 66*, 635–650.

Turing, A. (1936). On computable numbers, with an application to the Entscheidungsproblem. *Proceedings of the London Mathematical Society, 42*, 230–265.

Turing, A. (1948). Intelligent machinery. Technical report, National Physical Laboratory, London, reprinted in Ince, D. C. (ed.) *Collected Works of A. M. Turing: Mechanical Intelligence*, Edinburgh University Press, 1969.

Ulinski, P. S. (1990). The cerebral cortex of reptiles. In E. G. Jones & A. Peters (Eds.), *Cerebral cortex* (Vol. 8, pp. 139–215). New York: Plenum Press.

Vogt, C., & Vogt, O. (1919). Allgemeine Ergebnisse unserer Hirnforschung. *Journal Psychol Neurol, 25*, 279–461.

von Economo, C., Koskinas, G. N. (1925). *Die Cytoarchitektonik der Hirnrinde des erwachsenen Menschen*. Berlin: Springer.

Waldeyer-Hartz, H. W. G. (1891). Über einige neuere Forschungen im Febiete der Anatomie des Centralnervensystems. *Deutsche medicinische Wochenschrift, 17*, 1213–1218, 1244–1246, 1287–1289, 1331–1332, 1350–1356.

Waters, J., Schaefer, A., & Sakmann, B. (2005). Backpropagating action potentials in neurones: measurement, mechanisms and potential functions. *Progress in biophysics and molecular biology, 87*, 145–170.

Zakon, H. H. (2012). Adaptive evolution of voltage-gated sodium channels: The first 800 million years. *Proceedings of the Natural Academy of Science USA, 109*, 10619–10625.

Zhuo, M., & Hawkins, R. D. (1995). Long-term depression: A learning-related type of synaptic plasticity in the mammalian central nervous system. *Reviews in the Neurosciences, 6*, 259–277.

# Chapter 3
# Representational Mechanisms

**Abstract** This chapter attempts to move from the fundamental computational properties of the brain, previously described, into a sketch of how the brain builds a representation of the world. The next part of the book will deal more specifically with the linguistic portrayal we humans have of the world, a topic which has also served as a tentative approach to explaining the neural mechanisms that allow animals to build knowledge. In a sense, the core question of epistemology. Several philosophers, like Jerry Fodor (1983, 1990), have denied that an explanation of representations can be given in terms of neural biophysical properties. Today, such a radical refusal of the neurocomputational approach has become more and more marginal, in any case its thorough defense is beyond the scope of this book, and left to better advocates (Churchland 2002).

On the other hand, any endeavor toward explaining mental representations by neural mechanisms, has first to acknowledge that the notion of "representation" itself is problematic, and at the heart of current philsophical controversies. In addition, there are also positions entirely within a neuroscientific perspective, that deny the concept of representations completely. The title of this chapter leaves no doubt that we, instead, appeal to the notion of representations. It is out of the scope of this book the attempt to settle the philsophical debate on representations, and to lay down any new theory, our approach is to explore the emergence of semantic phenomena through the use of neurocomputational models based on a set of plausible mechanisms, that will be described in this chapter. Before that, we will offer a short overview of the philsophical issues posed by mental representation in general, with more detail on neural representation, and on neural computation over representations.

We will offer a selection of a small number of mechanisms, deemed to be at the core of the bridge between electrochemical activity and world representation. The ability of detecting coincidences, at different scale levels of neural circuits, is regarded as the most general, and probably most effective mechanism. Other more specific strategies will be discussed such as receptive fields, topological organizations, and selective processing pathways.

 
A. Plebe, V.M. De La Cruz, *Neurosemantics*,
Studies in Brain and Mind 10, DOI 10.1007/978-3-319-28552-8_3

## 3.1 Neural Representations

Intuitively, a quest for the existence of mental representations sounds paradoxical. We live embedded in our representations of the world, and of ourselves. Indeed, the initial enterprise of cognitive science was entirely dependent on the assumption of internal representations (Fodor 1975; Newell and Simon 1976), with little concern of the need to further explain what they were, and whether they exist at all. However, all attempts to provide a sound philosophical definition of mental representation turned out to be problematic. For Ramsey (2007) the notion of representation is a paradigmatic case of the well known Wittgenstein (1953)'s "family-resemblance" structure. The usage of the term representation among cognitive scientists (and philosophers as well) is highly variable, with radically different properties, often clarified with semi-formal, all-encompassing definitions, sometimes tapping into a general, pre-theoretical understanding of representation. Still, it is possible to appreciate in the diversity of uses of this term, a shared "family-resemblance". Even more drastic was Cummins (1989) in his discouraging the search for a definition of representation at all, claiming it as unfeasible in principle.

Next, we will discuss some of the most critical issues involved in defining representations, a comprehensive review is given by Ryder (2009a,b). As we will see, coming up with a definition of representation often requires formulating a comprehensive theory on how representations work. For the purpose of neurosemantics, the problem of representation is recast in the more precise terms of neural representations, discussed in Sect. 3.1.2. The close relation between representations and computations encourages the next theoretical quest, the one regarding the kind of computation that is performed by neurons. This, in turn, brings to the closing of this section, where the way biological neurons have of computing is contrasted with what ordinary computers do ordinarily, setting the stage for the modeling strategies that will be presented in the next chapter.

### *3.1.1 Representation and Its Troubles*

The troubles posed by mental representations are certainly not new in philosophy, dating back further than we would expect. According to Slezak (2002), most of the terms used in current debates were already central in the philosophy of the seventeenth century. For example, the theory of ideas of Nicolas Malebranche (1675), if stripped of its theological trappings, is much like a modern tripartite concept of representations, with what is represented, the representation, and the user of the representation. Likewise, the fierce critique leveled against this theory by Antoine Arnauld (1683), denying intermediate entities between perception and action, closely parallels the modern day enactivist anti-representation movement.

Probably the most serious trouble representations have is that of accounting for *misrepresentations* (Dretske 1986), for example, such as when we make mistakes in categorizing perceptual stimuli. Dretske (1981) articulated one of the most influential definitions of representation, based on causal relation. A mental concept can be described as a representation of an external entity, if this entity consistently causes the mental concept to be elicited. He recognized that this account has trouble in explaining why different entities may occasionally activate the same representation. For Slezak (2002) misrepresentation is just a different description of the well-known classical "Argument from Illusion", used to support sense-data as the immediate objects of perception (Ayer 1940): cases in which the correspondence between representations and the world fail.

Leaving historical recursions aside, misrepresentation is still problematic. The way out suggested by Dretske is by distinguishing a learning period, during which a subject learns that a certain representation refers to an entity type, and can make mistakes. This period fixes the reference, while the post-learning period leaves room for wild causes. This proposal is exposed to several objections, like the difficulty in identifying a sharp division between learning and post-learning periods. Even worse, it is vulnerable to additional troubles that resemble misrepresentations.

One is the classical problem of normativity. We have the clear intuition that in cases of misrepresentation something has gone "wrong", and even if a causal theory is able to describe different levels of accuracy of a representation, it has trouble in explaining why the misrepresentation is "wrong". A second, is the so called "*qua* problem" (Devitt 1981): a perceived stimulus, a cat for example, can cause several kinds of representations, in this case a cat but also a pet, or an animal. It is in the intention of the agent to think of *qua* as such-and-such. Probably, the most successful answer to this set of problems is teleosemantics. Millikan (1984) developed this idea introducing the concept of "proper function" of a biological trait, as its effect, the having of which is responsible for the continued reproduction of members of a population endowed with such a trait. In other words, the proper function of a trait is the property for which it has been selected, in an evolutionary sense. Applying this idea to representations, the content is now fixed by the class of stimuli the representational mechanism was selected for. Normativity is secured: cases of misrepresentation are "wrong" because the mechanism failed to perform its proper function. The *qua* problem is solved with the concept of "template", the kind of question one asks when tracing an entity, and this selects the appropriate representation (Millikan 2000). A similar teleological proposal has been developed by Papineau (1987, 1993), who appeals to the history of evolutionary selection as well, but assigns a privileged role to the content of desires. Belief contents should be explained in terms of desire contents, and desires have their content by representing what they are desires for.

Of course teleosemantics soon encountered its own set of problems, one of the most famous being the Swampman of Davidson (1987), an imaginary copy of himself generated by a really peculiar random aggregation of molecules. Swampman is physically identical to Davidson, but lacks any evolutionary history, therefore, none of his representations have any content, according to teleosemantics. Whether this, and several other objections succeed is a controversial matter, and beyond the scope of this section.

There are reasons, detailed in Sect. 3.1.3, that make teleosemantics scarcely relevant to our project. On the contrary, there are other less popular approaches, that are more congenial to our purposes. The focus of these approaches is on developmental history, rather than evolutionary history. We mentioned how Dretske already used the notion of a learning period in his causal theory, but without an adequate characterization of this notion. A refined elaboration on this is given by Prinz (2002) with the idea of "incipient causes". A representation has as content the class of entities to which the entities that caused the original creation of that representation belong. The remarkable difference is that for Dretske, during his learning-period, every entity that would cause a representation to be tokened gets included in its content, while for Prinz only those things that actually caused the first tokening of the representation, its "incipient causes", will count as content.

A different causal-developmental theory has been defended by Rupert (1999), based on statistical considerations. The content of a representation is defined as the class that elements in the past have caused to token that representation instead of another. Therefore, the actual content of a mental representation is determined by a substantive developmental process, a result of the physical disposition of the organism to represent, shaped by the individual's developmental interaction with the environment. The solution to the *qua* problem is inherent to the statistical definition of representation content: even if a cat is a pet and an animal, the past frequency that caused to token the representation "cat" is much higher than pets in general.

So far we have covered some of the philosophical problems of mental representations, and how various theories fare with these problems, under the tacit assumption that something like mental representations do exist, and have their physical instantiation. There are positions that are radically distant, that deny the existence of representations at all, or the possibility of their physical instantiation. One of the most authoritative is Dreyfus (2002), rooted in existential phenomenology, especially the works of Heidegger and Merleau-Ponty. We are not going to discuss these sorts of positions, because their attacks concern mainly the classical vision of representations as logical symbol structures, and tend to dissolve when moving to current neurocognitive accounts, as pointed by Grush and Mandik (2002). We will concentrate on challenges specific to neural representations in the next section.

### 3.1.2 Do Neurons Represent?

Much like the notion of representation in the mind is intuitively accepted, as has been discussed above, in neuroscience, representational vocabulary is used casually to characterize various neural processes. We are going to do the same, discussing in this chapter a series of brain mechanisms we identify as strategies for building representations. Thus, before examining how representation at the neural level fares with the troubles previously listed, a compelling question that arises is whether by construing certain brain mechanisms as representations, we are using a fictional, or even worse, a misguiding concept.

Doubts on the genuine nature of neural representations are raised by Ramsey (2007), who traced a distinction between *structural representations*, and simpler *receptor representations*. While the former maintains a structure of relations of the target domain being represented, the latter is simply the selectivity of the response to a class of external stimuli. According to Ramsey only structural representations, which are exclusive to the classical computational theory of cognition, are genuine notions of representation. Receptor representations, used in neural computation, fail to meet his "job description challenge", i.e. the level of explanation they give when construed as receptors do not change when construed as representations. However, as remarked by Sprevak (2011), it is far from clear why a receptor based notion cannot be explicative as representation, and Shagrir (2012) shows how Ramsey's concept of receptors does not meet with the usage of representations in current neural computation, where structural relations between neural signals and target patterns are common.

While the criticisms of Ramsey targeted neural computation, in favor of the classical computational theory of cognition, there is a strand within the field of cognitive neuroscience, that claims brain processes are best studied using noncomputational and nonrepresentational ideas and explanatory schemes. The tools of reference are those of dynamic systems theory, and one of the flagship arguments is the dynamical description given by Gelder (1995) of a mechanical device, the Watt governor. It is a clever device, invented by James Watt in 1788, following a suggestion from his business partner Matthew Boulton. This device controls the speed of an engine by regulating the amount of fuel, in order to maintain a near-constant speed. In the words of van Gelder

> It consisted of a vertical spindle geared into the main flywheel so that it rotated at a speed directly dependent on that of the flywheel itself. Attached to the spindle by hinges were two arms, and on the end of each arm was a metal ball. As the spindle turned, centrifugal force drove the balls outward and hence upwards. By a clever arrangement, this arm motion was linked directly to the throttle valve. The result was that as the speed of the main wheel increased, the arms raised, closing the valve and restricting the flow of steam; as the speed decreased, the arms fell, opening the valve and allowing more steam to flow. The engine adopted a constant speed, maintained with extraordinary swiftness and smoothness in the presence of large fluctuations in pressure and load.

In his essay, van Gelder gave a mathematical description of the Watt governor by differential equations of the arm angle and the speed of the engine. This system of equations completes the explanation of the system. An alternative, one preferred by a cognitive scientist, would be based on taking the arm angle as a *representation* for engine speed. Van Gelder argues that this idea is unwarranted for several reasons, the most compelling is that a description in representational terms does not explain anything over and above the explanation given by the dynamical equations. This example has been the focus of a wide debate, for Bechtel (1998) there are interpretations of the Watt governor, in which the representation analysis becomes pertinent, instead, according to Haselager et al. (2003) the current accounts of representation make it impossible and useless, to establish whether a system is representational or not.

The most fervent rebuttals of neural representations today, under the flag of dynamic system theory, come from the proponents of the enactive and embodied accounts of cognition (Chemero 2009; Hutto and Myin 2013). The most harmful threat to neural representation, according to Hutto and Myin, is what they call the "Hard Problem of Content", the idea that neural signals can only exhibit mere covariance relations with external sources of information, and covariance does not suffice for the existence of content, as for example, it cannot provide satisfaction conditions. By waving this weapon Hutto and Myin (2014) criticized the approach of Colombo (2014a), who showed how certain social behaviors, in which humans comply with norms, can rely on neural representations of beliefs and desires. A wrong conclusion, due to Colombo, falls into the Hard Problem of Content, leading Hutto and Myin to request "neural representations, no more pleas, please" (see also the reply of Colombo 2014b). The Hard Problem of Content is not so harmful for Miłkowski (2015b) who demonstrated how satisfaction conditions of content can be derived from neural representations in the case of anticipatory mechanisms of rat maze navigation. Notoriously, this task is supported by place cells in the hippocampus (O'Keefe and Nadel 1978), a discovery awarded with the Nobel prize in 2014. In the end, according to (Miłkowski 2015a), the Hard Problem of Content adds nothing to the list of troubles already analyzed in Sect. 3.1.1 for representation in general, like misrepresentation and normativity, has therefore already been solved, at least in part.

This short account does not intend to give justice to the debate on neural representation, and, as said before, in our book we are not taking up the burden of proving the philosophical consistency of a neural representation account. Our stance is to make use of this notion, adopting a weak theoretical commitment. Bechtel (2014) has shown how describing neural activity in terms of representation is a useful guiding principle, in the progress toward the mechanistic explanation of a phenomena, using the same case of place cells in the hippocampus mentioned by Miłkowski. To further legitimate the use of neural representation, we will now present several proposals that, unlike us, did take upon themselves the burden of proving how one can resist the philosophical troubles of representations.

### 3.1.3 Neural Representation Defended

The applicability of the notion of representation to neurons is defended by Mandik (2003) designing computer simulations of simple organisms, simulations of a kind that are very different from those presented in the second part of this book. The physical structure of the organisms is modeled using Framstick (Komosinski 2000), a collections of connected "sticks", and the control is based on computing elements loosely called "neurons". The simplest simulated behavior in a four legged Framstick creature is to coordinate its limbs so to walk in a straight line, more complex creatures perform vision driven taxis for food. Even if "neurons" of Framstick creatures have no biological plausibility, they fulfill the role of a conceptual demonstration of neural representation, since the contents of some of these elements are numerical values causally related with external objects, as in certain states of biological neurons. It is the case, for example, of the values carried by the neurons connected with the sensorial transducers of food location. Mandik has the further aim of comparing representations in terms of their different "economy": perceiving, storing memories, or commanding motors. We will leave this aspect aside, and take just the case of visual representations. The contents of the computing elements are genuine representations, according to Mandik, through a teleological account, in that locating a food source by sensorial neurons is a function in the survival economy of the creature.

This function is "proper", in the sense of Millikan, because it is fixed by the evolutionary history, simulated in Framstick by an evolutionary algorithm, that adapts the connection weights of the artificial neurons, using as fitness function the life span of the creature.

A different road is taken by Ryder (2004) in his SINBAD (*Set of INteracting BAckpropagating Dendrites*) theory, based on a theoretical speculation, grounded in neuroscientific evidence. The main claim is that talking about representations is appropriate for a specific class of brain neurons: cortical pyramidal cells (see Sect. 2.3.2). The peculiar behavior of these cells is the homeostatic contribution of their principal dendrites: they tend to contribute equally to the firing of the pyramidal neuron in the long run, but in order to achieve this stable state each dendrite needs to adapt its synaptic efficiency taking into account the probability of firing of the presynaptic cells at every other principal dendrite. It is feasible, notes Ryder, only if certain correlations hold between all the dendritic afferents, correlations reflecting regularities in the external world. For example, two dendrites may receive afferents from different sensorial features of the same object, like color and shape. This reconstruction of how a neuron becomes tuned to sources of correlation in the world is similar to our more general account of coincidence detection that will be described in Sect. 3.2.

Common to Mandik is the appeal to teleosemantics: Ryder claims that the working of the cells as described by SINBAD theory, work that way because they have been designed by evolution to fulfill the proper function of yielding reliable correlations of external sources, whose survival value is given by the predictive capabilities of what is represented by pyramidal cells.

A third defense of neural representation is offered by Nair-Collins (2013) in terms of *structural preservation*, a mathematical specification of class of preserving relations more strict than isomorphism and homomorphism, that can be applied to neural signals. The theory has technicalities that are not relevant for our purposes, in summary, neural activities are able to establish a mapping of relations of properties in the external world. Despite the difference of this theory to the proposal of Mandik and Ryder, the strategy for defending structural preservation as genuine representations is, again, teleosemantics. What makes neurons endowed with representations is that their states have the teleofunction of bearing certain correspondence relations, such that its doing so is adaptive for the organism of which that state is a part, and this function has been fixed by natural selection.

While we are sympathetic with the three defenses of neural representations just described, and their contribution to specific aspects of how representations are constructed, we doubt that the appeal to teleosemantics, in evolutionary terms, is of any real help. The impression is that since teleosemantics has gained today a high prestige in philosophical theories of mental content, the safer move in order to protect any neural representation theory from the long list of troubles listed in Sect. 3.1.1, is to attach evolutionary teleosemantics to it.

There are two different concerns with adopting evolutive teleosemantics. One is that it is affected by a serious philosophical weakness. It is based on the idea that a certain representation mechanism has been "selected for" in evolutionary history. But, as noted by Fodor (2008, p. 3):

> [...] 'adaptation for ...', 'selection for ...' and the like are themselves intensional contexts (just like 'belief that ...' and 'intention to ...'). [...] So the situation is this: either natural selection is a type of 'selection for ...' and is thus itself a kind of intensional process; or natural selection is a type of selection tout court, and therefore cannot distinguish between coextensive mental states.

But our major concern is methodological. Unlike synchronic or developmental accounts of neural representations, entrenched in a rich synergy with an enormous body of neuroscientific research, teleosemantics did not succeed in connecting with empirical relevant research. As far as we know, teleosemantics never engaged in confronting, for example, with current theories of brain or cortical evolution (see Sect. 2.3.1).

Turning towards a developmental account not only allows us to focus on the processes we deem more crucial in the construction of neural representation, it also grants access to a wealth of detailed mechanisms like those described in Chap. 2 (see also de Charms and Zador 2000). Even if our brain is, to a large extent, the result of evolution, narratives about hypothetical evolution of representations given by teleosemantics typically take the form Gould and Lewontin (1979) have called "just so stories". In addition, developmental-causal theories have their weapons too for fighting the philosophical troubles of representations, even if not as developed as the arguments of the teleosemantics community, we have discussed for example the theory of incipient causes by Prinz (2002) in Sect. 3.1.1.

Turning again to the simulation models of Mandik, a developmental history seems more appropriate and natural than evolutionary history in fixing the representational contents of artificial neurons. In fact, the evolutionary algorithm adapts the system by changing the weights modulating connections between neurons. But in the brain synaptic efficiency is adapted by development, certainly not by evolution. This objection does not hold for the SINBAD theory, which dictates that specific contents of pyramid cells are fixed during development, it is the general power of predicting by catching correlations that is supposed to be conferred by evolution. As noted by Usher (2004) the link to evolution appears superfluous, and adds nothing to the validity of the theory itself.

Eventually, let us mention a project that makes liberal use of the concept of neural representations dispensing with evolutionary teleosemantics, the NEF (*Neural Engineering Framework*) (Eliasmith and Anderson 2003; Eliasmith 2013), in which the typical representation unit is at the level of a *population*, rather than single neurons (see Sect. 4.3). The bridge between signals at the level of neural populations, and conceptual content, is provided by the idea of a *semantic pointer* (Blouw et al. 2015), a sort of compressed mathematical transformation of the multidimensional vector of activity in a population of neurons, into a new mathematical space, whose dimensions can carry cognitive meaning. The transformation is performed using circular convolution (Plate 2003).

### 3.1.4 Do Neurons Compute?

In cognitive science the notion of representation has typically been ancillary to that of computation. Every manipulation of representations, or derivation of an action from representations, is a computation. We just reached the conclusion that for neurons to represent is a useful and plausible construction, but does it immediately entail that neurons compute?

All throughout the previous chapter, we have not only assumed that the neuron computes, but also, that it is very likely the basic unit of brain computation. We did not refer to a precise definition of computation, using the term loosely as a purposeful manipulation of physical (mainly electrical) signals. It is time now to examine how correct our account of computation for the neuron is, and more broadly, the account of computation in our neurosemantic modeling approach, with respect to a philosophically sound definition of computation.

There are in fact several definitions of computation available, ranging from that of theoretical computer science (Turing 1936; Kleene 1936; Church 1941; Rice 1954), to that of pancomputation, the idea that the whole universe computes (Wolfram 2002). There are several lists of criteria for deciding if something computes or not, such as that given by von Neumann (1961) or the more recent and longer list by Smith (2002), an overview is found in Fresco (2014). For our purposes, we draw on the taxonomy proposed by Piccinini and Scarantino (2010) and Piccinini and Bahar (2013) of generic, digital, and analog computations. A crucial concept in their grouping is that of *vehicles*, on which computation is performed, defined as "entities or variables that can change state". A generic computation is a manipulation of vehicles according to rules that are sensitive to certain vehicle properties and, specifically, to differences between different portions of the vehicles, including their temporal changes. A fundamental constraint on vehicles is that they need to be "medium independent", their manipulating rules should obey differences between portions of the vehicles, along specified dimensions, but should be insensitive to any more concrete physical properties of the vehicles. Piccinini and Bahar give as counterexamples the processes of cooking and explosions, involving physical alterations of the medium being processed.

Thanks to the concept of vehicles, it is possible to give a stringent definition of digital computation, as a generic computation whose vehicles are strings of digits, which are defined as macroscopic states whose type can be reliably and unambiguously distinguished by the systems of other macroscopic types. Digit types should be finite in number. Strings are just concatenations of digits. We can immediately recognize that digital computers, as they are defined in theoretical computer science, fit into this definition, however, it is more general than that of a Turing-equivalent machine.

In analog computations, the vehicles are variables that can vary continuously over time, and the manipulating rules are in the format of differential equations. The best mathematical account of analog computers is that of Pour-El and Richards (1981). Analog computations depart from digital computation in several respects, one is that the computed values are always approximations of the exact values expected by the governing rules.

Where are neurons in this taxonomy? One may expect analog computations, since all the main neural processes, described in Sects. 2.1.2 and 2.2.2 could in principle, be described by differential equations on continuous variables. The point is that in the computational analog account of Pour-El, only a restricted class of differential equations are effectively computable, too narrow a range for neural biochemical processes. Note that it cannot be ruled out that neural processes might be included in a different, and more inclusive, theory of analog computation, however Pour-El is still the best candidate in the market of theoretical computation over continuous variables.

Piccinini and Bahar discuss at length the possibility of neurons being digital computation, clearly this interpretation is invited by the apparent on/off signaling of the neurons by action potentials, that led McCulloch and Pitts to the logic interpretation of the brain discussed in Sect. 5.1.3. As concluded there, and from all that has been presented in Sect. 2.1, it is evident that this is not the case. In particular, any attempt to reduce trains of action potentials to strings of digits, would fail to meet the finite constraint on the number of types, due to the continuous time of the events. One may argue that even if spike events are not synchronized, and thus placed over a continuous time dimension, their approximation within discrete time intervals might be valid for cognitive processes. However, as remarked by Piccinini and Bahar, this objection faces the problem of separating the cognitive level from the level of implementation, which is improper in principle when the project is to characterize computation at the implementation level. Thus, the conclusion is that neurons do compute, and belong to the class of general computation, not digital or analog computation.

We are not convinced that all vehicles relevant to neural computation can pass the medium independence criteria, Piccinini and Bahar, mention voltage changes in dendrites action potentials, neurotransmitters, and hormones, as possible vehicles manipulated by neural processes, but their analysis is limited to action potentials. It is the proper candidate: what matters in trains of action potentials for neural computation rules, is their amplitude in time, which can certainly be characterized in a medium-independent way.

Action potentials are clearly the largest overt activity of the brain, but the most important process for building representations is in the interaction between action potentials and plasticity. Limiting computation to the rules governing actions potentials on neural circuits with fixed connections, leaves out the entire construction of neural representations, i.e., the essence of neural intelligence.

Can the basic mechanisms of plasticity be characterized as medium independent vehicles? By definition, plasticity is a type of physical change of the medium: arborization, growth of synaptic terminations, changes in the number of channels, in the average number of vesicles, and so on (see Sect. 2.2.1). One may object that still, it would be possible to define vehicles able to abstract away the physical changes occurring during plasticity, preserving the property that all computational rules can be defined in their parts. We would not discard this possibility, certainly it would be a much more awkward job than in the case of action potentials, due to the chain of complex chemistry and genetics on which plasticity relies. Existing computational models of plasticity typically abstract away from physiological mechanisms, much more than models of neural electrical activity. In fact, while we have concrete examples of instantiation of action potential-like vehicles in artificial media, with the new brain-machine interfaces (Craver 2010; Schouenborg et al. 2011), nothing similar exists for the vehicles involved in plasticity.

Nevertheless, we would not deny that neurons compute, and belong to a type of computation that is different from the digital one. We believe that the processes performed by neurons are properly construed as computations, even if computations are performed on medium-dependent vehicles. The problem is philosophical. When relaxing the medium independence constraint, it is difficult to preserve the definition of generic computation from the risk of including the so-called "trivial" pancomputationalism. Contrary to the genuine belief of proper pancomputationalists, like Wolfram (2002), who elevate computation to the level of pervading the whole universe, the final goal of trivial pancomputationalism is to discredit computation, as an attribute that can easily be ascribed to any system. According to Searle (1990) the characterization of a process as computational is up to the observer, there is nothing intrinsic in the object that makes a physical process computational. Even more compellingly (Putnam 1988) argues that every ordinary open system, such as a boiling pan or a falling snowflake, can be proven to be equivalent to a Turing-equivalent machine. The introduction of the concept of medium-dependent vehicles was an elegant way to give generality to computation, behind digital computation, escaping trivial pancomputationalism. Perhaps neurons have been left out too.

A definition of computation that is appropriate for neurons is still on its way, much like most of the neurosemantic account given in this book. As stated by Piccinini and Scarantino (2010) "Unlike 'digital computation', which stands for a mathematical apparatus in search of applications, 'neural computation' is a label in search of a theory".

### 3.1.5 Neural Computation Inside Digital Computers

The previous excursus on different types of computation concerned biological neurons, and is pertinent for the next sections of this chapter, that deal with biological mechanisms of neural representation. In the next chapter, computation will move on, towards ways of modeling the mechanisms here described, by means of ordinary, concrete computers. There is no discussion on the fact that computers offer, of course, plain digital computation. What we would like to argue here, is that there is nothing odd in the divergent type of computation that exists between biological neurons, and the means used to model them.

The affinity between digital computers and mind computation was a concern of the traditional perspective on computational explanation in cognitive science, and has been the stronghold of the Computational Theory of Mind (Fodor 1975; Pylyshyn 1981; Johnson-Laird 1983). This is not a requirement for neural computation, however. There is no need to settle what feature of digital computers allows them to explain what neurons do. The business in which computers are involved, in this case, is more humble, modeling: to reproduce the behavior of a system of a different nature, which can also be a computational system itself, of the same or different computational type.

A crucial distinction now arises between computational models that just reproduce a behavior of an external system as an output, and those whose reproduction is realized in a way that *explains* the behavior itself (Piccinini 2007). Note that in a traditional account of scientific explanation, every computational model able to reproduce with enough accuracy the behavior of an external system under a variety of initial conditions, would count as an explanation, in that it has predictive abilities (Hempel 1965). More recently, it has been argued that a scientific explanation also requires the description of a *mechanism*, in general (Bechtel and Richardson 1993; Machamer et al. 2000), and particularly, in neuroscience (Craver 2007). In this context the term mechanism is used in a technical sense, it requires the identification of components of the system to be explained, the definition of the functions of each component, and the relations between functions in producing the set of system behaviors.

The concept of mechanism can be applied to computational models too, and becomes a possible criteria for ascertaining which models give explanation of the modeled system (Piccinini 2006, 2007). More specifically, Kaplan (2011) and Kaplan and Craver (2011) propose the 3M (*model-mechanism-mapping*) constraint on computational models with explanatory power:

> A model of a target phenomenon explains that phenomenon to the extent that (a) the variables in the model correspond to identifiable components, activities, and organizational features of the target mechanism that produces, maintains, or underlies the phenomenon, and (b) the (perhaps mathematical) dependencies posited among these (perhaps mathematical) variables in the model correspond to causal relations among the components of the target mechanism.

This constraint does not work as a logical binary condition, in fact, complete mechanistic models of neural behavior are unrealistic, it is perfectly compatible with incomplete models, where details are omitted either for reasons of computational tractability or because these details are still unknown. It is mostly a guiding principle: computational models whose design principle does not care about the structural correspondence between model components and external system components are merely predictive models, while in the case in which the design is based on such correspondence, they aim at being explanatory models. Most of the models developed for neurosemantics, described in the second part of the book, attempt to meet the 3M criteria, with different degrees of detail, and plausible correspondence of their components with brain components.

## 3.2 Coincidence Detection

In order to discuss what we deem to be the most important mechanism in neural representation, we employ the expression *coincidence detection*, as a general ability to detect coincidence in signals, and be affected by them. It is quite likely not supported by any single physiological mechanism, but is rather a broad brain phenomenon, that takes place at different scales of both the signals and the neural circuits involved. With increasing scale, we can break up coincidence detection into three levels, which are as follows:

1. two connected neurons level;
2. level of a neural population inside a single area;
3. level of two different brain areas.

The first level is precisely the one referred to by the principle of Hebb (1949), according to which a temporal coincidence in the activation of two connected neurons, if repeated for enough time, would result in an increase in the synaptic efficiency between the two neurons. This physiological modification captures the experience of the coincidence of two events, both in the case of external events for sensorial neurons, and of internal signals within the organism. In any case, the level of two neurons is at microscale, and applicable signals do not correspond directly to events as described in everyday experience (an exception would be the extreme case of code sparseness, see Sect. 4.3.1). This is the level where a sound understanding of the main processes has been achieved over the last 30 years. As discussed in Sect. 2.2.2, changes in synaptic efficacy between two neurons account for much more than pure Hebbian growth, sustained by a large amount of cellular and molecular mechanisms, nonetheless, the coincidence between signals is the main event captured by synaptic change. Note that we are not assuming that detecting coincidences is the final function to be reached by the system of two connected neurons, which will be dependent upon the specific patterns of inputs, the point is that in developing this final function repeated coincidences inside the two patterns play a crucial role. Moreover, we have seen how coincidence detection between two neurons comes in many forms, including the refined spike-timing-dependent plasticity, where the association is reinforced only when the presynaptic firing occurs before postsynaptic activation, within a few milliseconds.

The second level involves neurons at such a number that it allows us to compare their activity to an elementary cognitive event, such as a perceptual stimulus, and it was also originally conceived by Hebb, in terms of *cell assemblies*. This notion, though rather vague, makes sense in postulating that the coding of macroscopic experience should involve the coordinated activity of large numbers of neurons.

A good example of the interpretation of coincidence detection at the second level is the organization of circuits in sensorial cortical maps, those introduced in Sect. 3.3.3. One of the main features of these maps is the abundant lateral connectivity, as seen in Sect. 2.3.2, which can always be broken down into single synapses between two neurons, ruled by Hebbian plasticity, possibly extended by all its known variants. However, the signals are now characterized by a high degree of correlation, and the effect of coincidental signals, their self-sustainment, requires an interpretation at a more global level. A mathematical framework suitable for this interpretation is the one furnished by the theory of self-organization, which will be described in more detail in Sect. 4.1, and will be one of the bases in the neurocomputational approach to semantics here proposed. Here we anticipate just one particular phenomena, in order to show how well it fits within the general idea of coincidence detection. One of the first phenomena reproduced with the mathematics of self-organization has been the selectivity to orientation in the primary visual cortex (von der Malsburg 1973). It can be regarded as an instance of coincidence detection, the coincidence of a series of signals, being aligned along a specific direction in the retina. It is a learned feature, captured by repeated exposure to lines with similar orientation in the visual field. Different neural populations, inside the same cortical map, concur in detecting and gradually coding, different orientations, in various spots in the visual field. The detection of coincidence in the alignment of other patterns instead, is not arbitrary, in that in most cases lines belong to shape boundaries, one of the most precious bits of information in the vital task of segregating objects in the world.

A definition of this second level of coincidence detection is not limited to sensorial cortical areas. It can comprise the organization of neurons such as those named *neural cliques* by Tsien (2007). He identified possible neural cliques in mice who experienced startling episodes varying from the blowing of a puff of air, a short vertical freefall, or the shaking of the cage. A kind of "general startle" neural clique was identified, where cells respond to all types of startling stimuli, whereas more "specific startle" cliques collect neurons that respond to a combination of two types of, but not all, startling events.

The spatial scale of the events and the neural assemblies involved are not the only extended aspects compared to local Hebbian plasticity. The time window within which two events are perceived as coincident can be bigger than the few milliseconds proposed in the Hebbian case. In learning the association between taste and visceral distress, rats developed conditioned taste aversion even when the interval between the presentation of the tastant and the malaise-inducing agent was that of hours (Dudai 2007).

Eventually the third level concerns coincidences between patterns of activation in distant parts of the brain, not just in the same cortical or subcortical area, at a large scale dimension of signals, corresponding to events that are representative at a psychological level. Examples are the associations built between the amygdala and the visual cortex, or the auditory cortex, studied by LeDoux (2000). It is uncertain whether distant coincidence detection and coding is supported by the same local mechanisms, replicated over long chains of connections, or whether there is something else in addition. It has long been supposed that relations between patterns of activation of distant groups of neurons are conveyed by the synchronization between spikes (von der Malsburg 1995a; Singer 1995). It is interesting to note that it is again a kind of coincidence detection, but distinct and independent from the coincidence between neural signals interpreted as firing rates, as in classical Hebbian learning. Not only does it require very precise temporal resolution in encoding, transmitting and evaluating the temporal structure of stimuli, synchronization also introduces independent information, for example, two cells could be highly correlated, even if their firing rates are weak. Distant neural communication is still not confirmed on the whole, and the various hypotheses that have been put forth are controversial. For example, Fries (2005) holds that phase-locking among oscillations in distant neuronal groups allows communication between them, and the absence of neuronal coherence prevents communication, but that the phase (amount of synchronization) by itself is not a representational code. However, there is evidence that synchronization between hippocampal and parahippocampal regions modulates the encoding of events (Axmacher et al. 2006), and the neural representation of external locations is supplemented by phase information (O'Keefe and Recce 1993).

### *3.2.1 The Psychological Side of Coincidence Detection*

Coincidence detection, as we have put it forth, is an idea concerning how neural circuits come to represent facts and events of the world. However, it naturally shares resemblance with the psychological idea that one fundamental way of acquiring knowledge is by mental association. More broadly speaking, this idea is certainly not an innovative proposal of modern psychology, it can be traced back to Greek philosophy. It was with Aristotele (350BCE) that the concept of mental association was clearly enunciated. He acknowledged that associating at least two different experiences in contiguity, for a number of times, would lead to habit-formation and purposive thinking. Little attention was dedicated to this idea in philosophy, until the emergence of British empiricism, with Locke (1690), and especially Hume (1739), elaborating on the concept of association in detail. Among his basic principles, mental ideas become associated with one another through the experience of such things as spatial contiguity, temporal contiguity, similarity, and dissimilarity of sensations or simple ideas. In figuring out which correlations between two given events lead us to ascribe a causality relation, he identified the experience associations of *contiguity* and *contingency*. The former is the temporal and spatial proximity of events, the latter their regular covariation.

With the emergence of psychology as a separate and distinct field from philosophy, associative ideas become a central part of the theories on learning (Ebbinghaus 1885; Thorndike 1892), and find a strong empirical foundation with the studies of Pavlov (1927) on conditioning. In this form of association the two events are fully asymmetric: one is the *conditioned stimulus* (CS), which is neutral for the organism, the other is the *unconditioned stimulus* (US), of biological value. After the repeated experience of coincidental pairing of the two stimuli, the organism exhibits a conditioned response (CR) to the conditioned stimulus, even when it is presented alone. Pavlov's theories of conditioning were hugely influential during the behaviorist period in psychology, and especially in regards to viewing learning as a matter of strengthening or weakening connections between environmental stimuli and the behavioral response they evoked in organisms. Conversely, interest declined greatly as a result of the cognitive turn in psychology, and because of the general disregard for stimulus-response based theories that ensued. A new surge of interest was spawned by the attempt made by Rescorla and Wagner (1972) to lay the foundations for a mathematical formulation of conditioning. Their model is chiefly concerned with quantifying the strength $V$ of the connection between elements representing the two stimuli in Pavlovian conditioning. Assuming there is a variability of possible conditioned stimuli, belonging to a class $\mathscr{C}$, and a single unconditioned stimulus, each conditioned stimulus $c$ is associated with a strength $V_c$, which is updated at every new experience, by the following equation:

$$\Delta V_c = \alpha_c \eta \left( \bar{V} - \sum_{i \in \mathscr{C}} V_i \right) \tag{3.1}$$

where $\bar{V}$ is the asymptote of the associative strength, depending on the intrinsic value of the unconditioned stimulus, $\alpha_c$ is the specific salience of the conditioned stimulus $c$, and $\eta$ a learning rate. In case a stimulus $x$ is not followed by the unconditioned stimulus, the value of $\bar{V}$ is assumed zero, and from Eq. (3.1) $\Delta V_c$ become negative.

There are several difficulties when the abstract terms of Eq. (3.1) are projected onto the complexity of real learning experiences. Rarely can the actual stimuli humans and animals are presented with be understood as atomic entities, more often than not, they are made up of several distinct components, which can sometimes but not always share the same unconditioned stimulus. This problem has two contending proposed solutions: the *elemental* strategy in which the components can be perceived as unitary configurations, or the *configural*, for which the components maintain their individuality also when appearing in conjunction (Williams et al. 1994).

However, the Rescorla and Wagner model has become a point of reference for many later developments in the psychology of associative learning, which are not addressed here. This field has a strong familiar resemblance with our coincidence detection proposal, but there are notable theoretical divergences, in addition to the obvious differences between psychological and neurocomputational levels, mentioned before. In psychology, associative learning is mainly considered to be one among several independent possible forms of learning. Typically, for psychologists, nonassociative forms of learning include phenomena like habituation, priming, and perceptual learning (Hall 1991), where no explicit contingencies between the stimuli to be learned or actions are observed. On the other hand, coincidence detection is conceived as the more basic mechanism by which a neural system gains internal representations of events and facts of the external world. It may be perfectly applicable to phenomena classified in psychology as nonassociative learning, in which there is no overt definition of more than one stimulus, and all structural contingencies amidst the elements of a stimulus, that inevitably exist, are neglected in the relevant psychological theories. Those structural contingencies are the most likely cues for coincidence detection and subsequent coding. More about coincidence detection and structures will be discussed in Sect. 3.2.2.

It is not surprising that associative learning finds more than one strong opponent within psychology and cognitive science. One of the most radical confutations of associationism in general comes from the Computational Theory of Mind, a project construed precisely as its alternative, since its foundation (Fodor 1987, p. 18):

> Exactly what was wrong with Associationism, for example, was that there proved to be no way to get a rational mental life to emerge from the sorts of causal relations among thoughts that the 'laws of association' recognized. [...] Here, in barest outline, is how the new story is supposed to go: You connect the causal properties of a symbol with its semantic properties *via its syntax*.

The same position is still held today (Fodor and Pylyshyn 2015, p. 10):

> we aren't associationists; that is, we think that mental processes are typically causal interactions among mental representations, but not that such interactions are typically governed by the putative laws of association. To the best of our knowledge, embracing RTM is the only way that a naturalist in cognitive science can manage to avoid associationism and/or behaviorism, both of which we take to be patently untenable.

This controversy is a contemporary continuation of the enduring opposition between a rationalist perspective, taken by Fodor and Pylyshyn, and empiricism, a debate by large outside the scope of this book. A defense of empiricism, that shows ways of explaining causal properties of concepts acquired by associative laws, denied by Fodor, is found for example in Prinz (2002).

One of the most radical and detailed arguments against associative learning is marshaled by Gallistel (2000, 2010) and Gallistel and Balsam (2014), drawing as much evidence as possible from the literature of psychology and neuroscience, that cannot be sufficiently explained by associative learning. His preferred examples bear on invertebrate navigation, like the ability of desert ants to forage as far as 50 meters away from their nests, in random directions, and then get back along the shortest path, or the use of solar ephemeris by bees. Other alleged counterexamples of associative learning include learning from few isolated experiences,[1] or cases where the timing between the conditioned and the unconditioned stimuli is not standard.

The conclusion is that instead of a general purpose associative-like learning, our brain is equipped with a series of innate specialized learning modules, and that what is learned is in the form of symbols. Several challenges raised by Gallistel to associative learning in its standard formulation are important and stimulating, several details have been addressed within an associative account, like the temporal separation of the stimuli (Dylla et al. 2013), details we cannot get into here. What we highlight here however, is the resemblance between the psychological theory of associative learning and our proposed coincidence detection. But the latter, as we put it forth, is a different notion, it is an idea concerning how neural circuits come to represent facts and events of the world. Nevertheless, the general dismissal of associative-like learning by Gallistel, in favor of innate specialized modules that learn symbols is deadly for coincidence detection too. Our opinion is that the alternatives suggested by Gallistel fair much worse than the associative accounts in matching brain processes. He is well aware of this issue (Gallistel and Balsam 2014, p. 141):

> Perhaps the biggest obstacle to neurobiologists' acceptance of the view that the brain stores information in symbolic form just as a computer does,is our inability to imagine what this story might look like at the cellular and molecular level.

We cannot tell whether it is just a matter of lack of imagination. Surely, what is lacking so far, is evidence for a neural mechanism for storing information in symbolic form, against a wealth of evidence for mechanisms compatible with associative learning. As far as innate specialized modules are concerned, while the examples given for invertebrates are quite compelling, much less is provided by Gallistel for humans. This is his best example (Gallistel 2010, p. 574):

> Chomsky's suggestion grew out of his recognition that learning was a computational problem – a view that is foreign to the associative conception of learning [. . .] and to most neurobiological conceptions of learning. [. . .] The example Chomsky had foremost in mind was the learning of a language. [. . .] The computational challenge this poses is so formidable that there is no hope of surmounting it without a task-specific learning organ, a computational organ with a structure tailored to the demands of this particular domain.

Unfortunately this example, as we will see in Sect. 5.2.1, did not withstand the scrutiny of neurophysiological plausibility.

On the other hand, it should be said that in psychology we can also find positions that give credit to a wider and more general role of associative learning (Shanks 1995). More recently Heyes (2012) has disputed the *association-blindness* that cognitive science has induced in comparative cognition studies, or the tendency

[1]One of the models in this book (Sect. 6.2.3) demonstrates the feasibility of learning from a limited number of experiences, using coincidence detection only.

of refuting associative explanations, as oversimplifications in explaining complex intelligent behaviors in animals. Even in social cognition, an area in which the more sophisticated forms of behavior are studied, there are important examples that can be explained by associative learning. Wunderlich et al. (2011) using fMRI studied subjects performing a renewable energy management game. In the area with the highest correlation with the game parameters, the right midinsula activation values best fit with a model of simple statistical covariations with the simulated environmental parameters of the game. In a *two-armed bandit* game, Burke et al. (2010) have shown that the main strategy used by gamblers is based on observational associations. In a similar experiment, with the addition of suggestions concerning the trust that should be assigned to future advice from an unknown confederate, Behrens et al. (2008) ascertained the associative learning of the degree of trustfulness of the confederates.

In understanding our concept of causation, a theory that has gained recent attention in philosophy is that of constant conjunction, based on the idea that causal statements are empirical, and are derived from our past experience by observing recurrent coincidences of conjunction between objects (Liu and Wen 2013). This view is perfectly in line with the account given by Churchland (2010) on the Hebbian way of learning causality in the case of prototypes, that is closely related with our coincidence detection principle.

### *3.2.2 Coincidences and Structures*

Our notion of coincidence detection accords well with certain aspects of *structural* theories of neural representation, which in turn share the general idea of founding mental representation upon similarities between a conceptual structure and the structural properties of the referents. The appeal to sorts of similarities between representation and what is represented is certainly not the latest fashion in philosophy and cognitive science. It can be found as early as in the words of Aristotle (335–323BCE, 16a3):

> Now spoken sounds are symbols of affections in the soul, and written marks symbols of spoken sounds. And just as written marks are not the same for all men, neither are spoken sounds. But what these are in the first place – affections of the soul – are the same for all; and what these affections are likenesses of – actual things – are also the same.

On a similar vein, Hume (1748, ch.IX) asserts that

> All our reasonings concerning matter of fact are founded on a species of analogy, which leads us to expect from any cause the same events, which we have observed to result from similar causes. [...] But where the objects have not so exact a similarity, the analogy is less perfect, and the inference is less conclusive; though still it has some force, in proportion to the degree of similarity and resemblance.

The informal accounts of similarity and resemblance gained a first rigorous formulation in mathematical terms as structural isomorphism in the work of Russell (1927), and made their entrance in cognitive science with Palmer (1978).

Despite its advantages, first and foremost the ability to incorporate a theory of ontogenetic acquisition of concepts, similarity resemblance alone is modest ground for a theory of mental representations. It easily suffers most of the same troubles representation theories suffer from as seen in Sect. 3.1.1. More specifically, it fell prey to the criticisms raised by Goodman (1976), such as the symmetry property of structural isomorphism, that one does not expect in representations.

However, a different way of conceiving a resemblance has been proposed in the past decades, which is much more promising and compatible with neural mechanisms. It was first described by Shepard and Chipman (1970) as "second-order isomorphism". According to this view, a representation system does its job not because of the physical similarities between its vehicles and the properties of the represented system, but because instead the physical relations between its vehicles support a structural-preserving mapping between the two. Second-order isomorphism was elaborated by Edelman (1999) in a computational framework called "chorus of prototypes", as a theory of the visual recognition and representation of objects. The "chorus" is a high dimensional space of similarity whose orthogonal basis is a set of prototypical shapes. Swoyer (1991) provided a formal definition of structural similarities in abstract set-theoretic terms, and introduced the expression "surrogative reasoning" for the mental ability to process a structural representation in order to draw inferences about what it represents.

The proposal of a naturalized structuralist theory of representation has been put forth by O'Brien and Opie (2004, p. 14): "We will say that one system *structurally resembles* another when the physical relations among the objects that comprise the first preserve some aspects of the relational organization of the objects that comprise the second." They acknowledge the lack of understanding of the brain that would make it possible to identify the structural properties and consequent resemblance relations that might ground mental representation, however, they point at patterns of neural activities in connectionist networks as abstractions exhibiting structural resemblance able to ground mental representations. Recently Nair-Collins (2013) offered a mathematical specification of second-order similarities oriented towards neural representations, as *structural preservation*, however, to secure his theory he adds teleosemantics as well, as seen in Sect. 3.1.3. This is not our concern here, our commitment to mental representation theories has been discussed in Sect. 3.1.

First, we should note that all characterizations of structural similarity using mathematical set theory, even if helpful in formalizing theories, are exposed to the criticism raised by van Fraassen (2008): there's simply no sense to be made of the idea that a homomorphism, and even worse an isomorphism, might hold from one concrete, physical system to another, since the technical notion of this sort of relations is only well-defined in the domain of abstract, mathematical systems. O'Brien and Opie (2004) correctly use the notion of "structural resemblance", which is weaker than mathematical set theory relations. Similarities may be treated mathematically in a more plausible way in probabilistic terms, see for example our characterization of population coding in Sect. 4.3.2. More importantly, independent of how structural similarities have been formalized, the basic concept is that what is captured by neural vehicles is some kind of relation between structural elements of external objects or facts. It often implies that tokens of an entity present at least pairs of properties which are consistently part of the same entity. It is this coincidental relation that is captured by the neural system. For example, common to most tokens of "bicycle" is the simultaneous presence of two circular equal or nearly equal elements, with almost parallel axes of rotation. This is a second-order similarity, corresponding to a coincidence, over other relational similarities at a lower level, and corresponding coincidences. For that in each of the two circular elements (the bicycle's wheels), the structural property of edge points continuously changing their orientation up to 360°, holds. Thus, the gap of knowledge on the neural mechanisms supporting resemblance relations, bewailed by O'Brien and Opie, can be, at least in part, filled by the coincidence detection mechanism.

### *3.2.3 Simulative Representations*

Coincidence detection is probably one of the most powerful qualities of neural systems, at different levels, and can therefore assume very different forms and functions. Some are so peculiar that they deserve a specific and detailed account, even if they do not entail a completely new mechanism in action. This is the case, we believe, of *simulative* representations, where the coding of a perceived action is a sort of internal simulation of the action itself.

Aspects of neural simulation have been observed for some time (Ito 1984), and a few speculative theories have been proposed, such as the analogy with emulators in control systems (Grush 1997). However, only in recent decades have they gained widespread attention, and in particular, since the discovery of mirror neurons by Rizzolatti et al. (1988). Notoriously, the discovery sprung up unexpectedly, during direct measures of motor neurons in monkeys engaged in action tasks. During an interval of the experiment, it happened that a neuron, still being measured, was signaling activity, even if the monkey was sitting still, just observing others grasping food. Looking more deeply into this amazing observation, it turned out that about 20 % of neurons in the rostral part of the monkey ventral premotor cortex, Brodmann area 6, fired not only during normal action execution, but also when the monkey observed an action executed by another individual. This area was called F5, and also includes a different category of visuomotor neurons, called *canonical*, which also respond to the presentation of certain objects, which are often objects that are compatible with the specific action they code.

This discovery gave rise to a number of questions, first, the one concerning what the origin of the visual information was, since the F5 area lacks direct connections with the visual processing pathway. The best candidate was STS (Superior Temporal Sulcus), whose anterior part includes visual sensitive cells. Investigations along this path succeeded in identifying a second area, inside the infero-parietal cortex, with a population of neurons similar to those in F5, called PF (Rizzolatti et al. 2001). Among all visually responsive neurons in PF, 40 % were sensitive to actions, and of these, 70 % were also active when performing the same actions. Of great interest, of course, was establishing whether these peculiar neurons were also found in human brains as well. The limited possibilities of non invasive methods, compared to direct electrode measures in monkeys, limited the investigations, which came up with contradictory results. Studies on imitation using fMRI seem to give indirect confirmation of the existence of mirror neurons in humans (Iacoboni et al. 1999), but other studies based on cortical adaptation tend to exclude it (Lingnau et al. 2009). Currently the amount of positive evidence seems to confirm, almost unambiguously, the presence of mirror neurons in at least two regions: the inferior section of the precentral gyrus, plus the posterior part of the inferior frontal gyrus; and the inferior parietal lobule, including the cortex located inside the intraparietal sulcus (Rizzolatti and Sinigaglia 2010).

The confidence that mirror neurons exist in the human brain has ignited intense interest in ascertaining their possible role in a range of higher cognitive functions. It has been speculated that these neurons not only code elementary motor sequences, but that they even represent action intentions. Many experiments have been designed in order to verify a similar idea. For example, mirror neurons do not respond if the action does not reach an object, however, they fire if the destination point of the action is hidden, and therefore, an object is presumed (Umiltà et al. 2001), and it has been found that the intrinsic value of the object modulates their response (Caggiano et al. 2012). Nevertheless, whether the indirect evidence gathered so far sustains the conclusions that mirror neurons are really involved in action understanding is highly controversial (Hickok 2009). One of the most ambitious expectations is for mirror neurons to reveal the more hidden mysteries of human language (Rizzolatti and Arbib 1998). The capability of representing other's actions is held as the initial trigger for communicating, starting from the speech level: an individual can capture the phonetic form of an utterance, because she is observing her speaker, and this observation gets mapped onto her correspondent phonetic motor code. Even further, Glenberg and Gallese (2012) postulate that mirror neurons can provide the key to answering almost all aspects of language, from comprehension to production, from syntax to semantics. Indeed, evidence has been gathered on the involvement of visuomotor neurons in certain aspects of language. Pulvermüller and Fadiga (2010) review strong evidence of the interaction between the frontocentral brain action systems and the comprehension of phonemes, semantic categories and grammar. However, studies on left-damaged patients reveal a double dissociation between the ability to imitate pantomimes and the ability to produce and comprehend the corresponding action verbs, suggesting that processing action words is independent of the ability to produce the associated object-directed actions (Papeo et al. 2010). On the phonetic side, Hickok (2010) notes how the mirror neurons hypothesis corresponds to the old motor theory of speech perception, which was already ruled out by numerous demonstrations of the relative independence of the motor and perceptual speech systems. A recent study on patients with lesions involving motor regions found almost intact speech perception abilities, disrupted instead when lesions reached auditory regions in the temporal lobe (Rogalsky et al. 2011). On the pragmatic side, Toni et al. (2008) contend the possibility of bare action recognition in building up the foundation of communicating. Tettamanti and Moro (2012) call for caution in the attempt of relating syntax with mirror neurons, since syntactic structures are not directly available to visual or acoustic observations.

Independently from these intense discussions, our main point here is that mirror neurons are just one case – a notable case indeed – of the general coincidence detection principle for neural representation. As pointed out by Heyes (2010), a convincing sequence of the ontogeny of mirror effects might be the following:

1. at birth visual responding neurons in STS project weakly and randomly into the premotor areas;
2. all events during which an executed action is observed at the same time, for example at the mirror, imitated by parents or siblings, or executed in a group, induce a coincidental activation of neurons in STS and premotor cortex, reinforcing the reciprocal connections;
3. gradually the connections become so stable that only activation in STS are sufficient to elicit responses in mirror areas, whose connections are now coding for the seen action.

This is not the only possible account of the origin of mirror neurons, and several of the initial investigators implicitly endorse a specific nature of these type of neurons, genetically evolved, even if there is no direct evidence so far, a related discussion can be found in de Klerk et al. (2014). Nevertheless, empirical observations of imitation events, like those described in the first point, are abundant in infancy (Ray and Heyes 2011; de Klerk et al. 2014), and mirror neurons have been confirmed as being plastic to associative learning (Calvo-Merino et al. 2006). A sketch for a computational model explaining the development of mirror effects by Hebbian learning between STS, PF, and F5 was given by Keysers and Perrett (2004), a fully developed model by Cooper et al. (2013) contends the need of a "non-Hebbian" additional component, to account for the contingency between observed actions and executions of the same. This distinction is not meaningful for the general mechanism here proposed. Coincidence detection is not limited to the Hebbian principle, as in its original proposal, it is the ability of neural circuits to capture coincidental events, taking into account all relevant covariations. Therefore, it includes the decrease of connection strength between two close events, in the case of the likelihood of the second, in absence of the first, and it is all that is required for taking into account contingency as well as contiguity. As mentioned in Sect. 2.2.2, it is argued that spike-timing-dependent plasticity can be a valid candidate for this job, by combining Hebbian or anti-Hebbian effects, depending on which action potential occurs first, the presynaptic in the first case, the postsynaptic in the second. Despite several indirect types of evidence, the role of spike-timing-dependent plasticity is still a matter of debate (Feldman 2012; Koch et al. 2013). In any case, the discrimination between contingency and contiguity need not be solved at the direct level of a single synapse. World-level events always involve large circuits of neurons. If the sequence of two events is not characterized by contingency, the decrease of connection strength between neurons carrying information between the two events will take place naturally, due to occurrences of the second event in absence of the first, by synaptic depotentiation.

In conclusion, we would like to mention other principles that have been proposed as the basic workings of the brain, and could therefore be considered as rivals to coincidence detection. One principle that gained popularity in the last decade is known as "Bayesian" or "probabilistic" (Griffiths et al. 2010; Tenenbaum et al. 2011), which, quite obviously, dictates that the brain implements Bayesian inference. For example, when perceiving a mental representation $r$, the probability that it has been caused by an object $o$ is given by the conditional probability $p(o|r)$. The best job the brain could do is to believe that the actual object, among the class $\mathcal{O}$ of all possible objects that can cause $r$, is the one, $\tilde{o}$, that maximizes the conditional probability:

$$\tilde{o} = \arg\max_{o \in \mathcal{O}} (p(o|r)) \tag{3.2}$$

Unfortunately $p(o|r)$ are unknown, so here the help of Bayes theorem comes in:

$$p(o|r) = \frac{p(r|o)p(o)}{p(r)} \tag{3.3}$$

All probabilities on the right of Eq. (3.3) can be learned by experience, or their approximation.

The radical difference between our coincidence detection and the Bayesian hypothesis is that the latter is aimed at a purely phenomenological description of mental/brain behavior. The brain of course does not incorporate Bayes theorem, and proponents of this view do not claim that it does. On the contrary, coincidence detection is proposed as a mechanism of how neural circuits work. There have been a few attempts to derive hypothetical neural computations in a top-down fashion, that perform something similar to the abstract notion of Bayesian inference, by combining a coincidence detector with a mixture of Markov chains for example (George and Hawkins 2009), or by combining population coding with a winner-take-all layer (Nessler et al. 2013). We are taking another path, conceiving coincidence detection as here described, starting from the basic computational facts of neurobiology, and how they provide the ground for capturing facts of the external world.

## 3.3 Columns, Fields, and Maps

Coincidence detection is a common feature of the brain as a whole. In this section we switch back to a special part of the brain, the cortex, and discuss specific mechanisms, which may be behind its impressive power and aptness in representing meaning. The anatomical and cytological properties of this thin layer of neurons have been described in Sect. 2.3. The next subsection is a kind of bridge between that view and a more representational oriented perspective. It deals with the cortical column, which is first and foremost, a vertical organization of neural circuits, but also prone to representation interpretations. The other subsections will discuss more basic representational mechanisms, the concepts of receptive fields, and of topological maps.

### 3.3.1 Columns

The word "column" (*Säule*) was first used by Constantin von Economo and Koskinas (1925) to describe cords of cells in the cortex oriented radially. The idea that these tiny cylinders might be the elementary processing unit was instead initially suggested by Rafael Lorente de Nó (1938). It was Vernon Mountcastle (1957) however, who discovered and demonstrated the concept of columns, from his observations of the somatic sensory cortex in cats and monkeys. He pointed out the double evidence for columns: on one side, the anatomical aspect, with the vertical cylinders of neurons separated by cell-poor neuropil zones every 30–50 m; on the other side, the functional relationship with all neurons in the column responding to stimulation of cutaneous receptors located at a particular site. The diameters of these cylinders match with that of the uniform neuron counts by Rockel et al. (1980), already discussed in Sect. 2.3.2. Further evidence came a few years later with the extensive studies of Hubel and Wiesel (1959), showing columnar organization in the primary visual cortex.

If the neuron is the basic element of brain computation, the cortex has its own more powerful computational unit, made up by a collection of cell phenotypes strongly and reciprocally connected: the column. As the importance of a single neuron has prompted for the development of computational models of its basic behavior, similarly, the evidence for the column as the functional element of the cortex has given rise to the search of a unified model, able to explain the most essential functional properties of columns, wherever in the many areas of the cortex they may be.

The first attempt in this direction was made by Marr (1970), who proposed a "fundamental neural model" of cortical columns. His approach was to start from a mathematical formulation of the general problem of classifying input signals, developing an equivalent representation using neuron-like units, and then fitting the sketched model with the variety of cells in a cortical column. This attempt was both too ambitious and too distant from empirical reality, and as a result was almost totally neglected. Nearly two decades later Shepherd (1988) proposed a model, which was at the same time much simpler but more closely related to the physiology of the cortex, based on an abstraction of the cortical pyramidal neuron as an integrator of all excitatory inputs at the spines of its dendritic branches, further modulated by the excitatory and inhibitory inputs along the apical shaft and into the soma. Two of these essential pyramidal neurons are arranged as superficial and deep representatives, together with a spiny stellate excitatory and two inhibitory essential neurons. Independently, Douglas et al. (1989) proposed a model that was

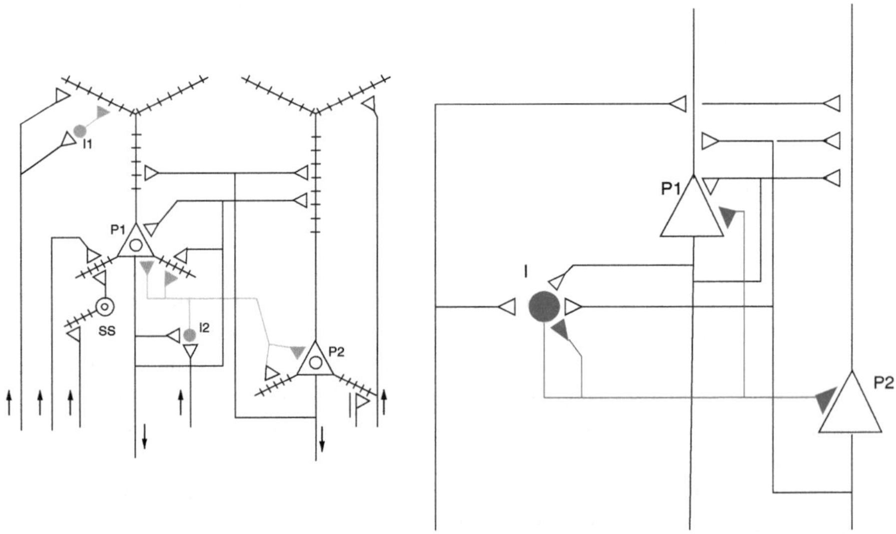


**Fig. 3.1** Diagrams of possible "canonical" circuits in the cortex. On *the left* the canonical neocortical circuit as represented by Shepherd (1988), with P1 and P2 the superficial and deep pyramidal cells, a spiny stellate cell SS, and two inhibitory interneurons I1 and I2. On *the right*, the scheme of Douglas et al. (1989), where P1 and P2 are the superficial and deep pyramidal cells, and I an inhibitory GABAergic neuron

quite similar to Shepherd's, and called it the "canonical microcircuit" equivalent of the column. It is made up by the combination of three abstract neurons, two excitatory and one inhibitory, with each conceived as the average contribution of a larger number of cells in the column that belong to the same class of neuron. One of the two excitatory virtual cells represents the deep pyramidal population (of layers V and VI), the other represents the superficial pyramidal population (layers II and III) together with the spiny stellate cells of layer IV. These circuits are shown in Fig. 3.1.

The columnar organization of the cortex has been one of the most influential concepts in neuroscience, but as Mountcastle (2003) himself underlined, his discovery was met with disbelief by many neuroanatomists, and the ensuing 50 years or so that had passed since, had by no means shown it to be generally accepted. One of the most direct challenges raised has been that of Horton and Adams (2005), based on the argument that there are too many groups of vertical cells that act as single processing units in various areas of the cortex, with an impressive difference in diameter. In order to account for the diversity of these vertical cylinders a rather confusing nomenclature has been created, with "microcolumns" the original unit 30 m wide devised by Mountcastle, "classical columns" the cytochrome oxidase blobs patterns, with a diameter 10 times larger, observed by Hubel and Wiesel (1959) in the primary visual cortex, and "hypercolumns", the units covering a complete orientation set for a single retinal site in in the primary visual cortex, about 1 mm wide (Hubel and Wiesel 1974a). More recently Rakic (2008) insisted with the ambiguity in the criteria used to define a "column" and subsequent incompatible sizes, arguing that there is only one flavor of columns with a precise definition, his "ontogenetic column" that refers to the cohorts of cortical neurons that originate from a single neuronal progenitor (see his radial unit hypothesis in Sect. 2.3.1). However, the relationship between the ontogenetic column and the functional definitions of larger columns is not clear.

Details of this ongoing debate are beyond the scope of the present work, but for an in-depth discussion see Nieuwenhuys et al. (2008, pp. 586–590). For the purpose of grasping the essential computation common to all areas of the cortex, we sympathize with Maçarico da Costa and Martin (2010) in reproposing a kind of canonical circuit, that abstracts from the precise structure of the column. The exact type of circuit we would prefer in order to understand the semantic processes of the cortex will be detailed in Sect. 4.2.

### *3.3.2 Receptive Fields*

One of the most powerful concepts in relating activity in the cortex with the perceived external world is that of "receptive field". Its first use did not address the cortex, and was primarily used in the context of vision. It was introduced by Keffer Hartline (1938) as the area in the retina, which must be illuminated in order to obtain a response in a given neuron.

Yet, as early as 1928, Hartline was able to exploit the technology of single cell readings (Adrian and Matthews 1927a,b) in vision, by examining a very suitable animal, the xiphosuran arachnoid (*Limulus polyphemus*), commonly called "horseshoe crab", which lacks cortex entirely. Its lateral compound eyes are coarsely faceted, and receptor cells project to the brain by long optic nerves, in which single axons can be separated rather easily. The relation between the eye stimulus and the neural discharge is relatively simple, with each ommatidium having its own single neuron. Illuminating a single ommatidium, therefore, elicits firing of its connected neuron. The case of the *Limulus* however, turned out to be not so simple after all. When neighbor ommatidia are also illuminated, the discharge decreases, revealing inhibitory interactions, an intriguing effect that Hartline went back to study further several years later (1967). The need for an idea like that of receptive fields become necessary when Hartline, in 1938, after his initial success with the *Limulus*, undertook the same single axon analysis of the more complex optic responses of the retina in cold-blooded vertebrates. When recording from single axons Hartline found other behaviors, in addition to discharges similar to those in the *Limulus*, where there was firing for the duration of the light stimulus. What he found was activity appearing when a light stimulus was withdrawn, as well as activity correlated to the onset and cessation of illumination. Moreover, he was able to define the precise configuration of a receptive field, by charting the boundaries of an area over which a spot of light sets off impulses in a ganglion cell's axon.

His results were replicated in mammals by Stephen Kuffler (1953), who refined the definition of receptive field, by differentiating its anatomical and functional meaning. The anatomical configuration of a receptive field is the pathway of all receptors actually connected to ganglion cells, and is fixed at a certain stage of maturation of the organism. The functional meaning includes not only the areas from which responses can actually be elicited by retinal illumination, but also all those areas which show a functional connection, by an inhibitory or excitatory effect on a ganglion cell. In this respect, the field size may change depending on the illumination pattern, involving areas which are not in the immediate neighborhood of the ganglion cell and that by themselves do not induce discharges.

It is thanks to Hubel and Wiesel (1959) that the concept of receptive fields moved from neurons in the retina up to columns in the cortex, having discovered the now well-known selectivity to line orientation in the primary visual area. Their studies increasingly spread the double use of the receptive field concept: taken to mean the definition of an area on the retina that excites a column, or the specific properties of the input pattern that evokes the strongest activity in the column. This last use of receptive field is, for example, the one relevant for Hubel and Wiesel (1962) in the differentiation of columns in the striate cortex as "simple" or "complex", with the former maximally excited by the largest summation of light in its excitatory subfield, and no light in its inhibitory subdivision.

The focus on the shape of receptive fields, and the new picture given by Hubel and Wiesel, stimulated research efforts to find mathematical formulations that could characterize receptive fields in a concise and readable form. Examples are the difference of Gaussians for ganglion cells and neurons in LGN (*Lateral Geniculate Nucleus*) (Rodieck 1965; Rose 1979), or Gabor functions (Gabor 1946) for simple cells in V1 (Daugman 1980, 1985). A sort of "evolution" in how the notion of receptive field has been used and interpreted is given in Fig. 3.2.

It is interesting to point out that the concept of receptive field is not only of practical use in characterizing the specific behavior of cells in visual systems, it is first and foremost, a basic bridge between the electrical phenomena of cortical column firing, and the entity in the external world that caused the firing.

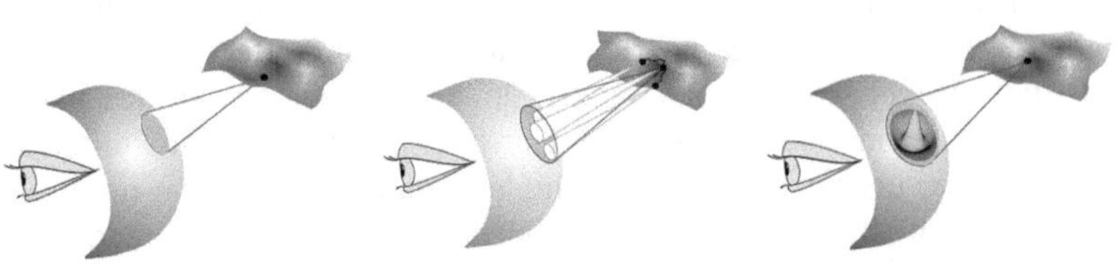

**Fig. 3.2** The evolution in the meaning of the expression "receptive field", illustrated by a sketch of the retina and a cortical neuron. On *the left*, the receptive field as originally introduced by Hartline, in *the middle*, including lateral connections, on *the right*, accounting for the shape of the field function

When the receptive field concerns ganglion cells, or thalamic cells in LGN, the relationship becomes relatively simple: the neuronal activity signals a specific sensorial experience, that takes place in a narrow area of the retina. A direct causal connection, of a topological nature, between facts in the external world and neural behavior can be established. Moving into the cortex, the receptive field of columns in V1 can still be a good explanation of the contents: the peculiar shape of objects in the external world on which columns are tuned, together with the topological constraints of where in the retina the stimulus of this object is projected. As soon as areas in the visual cortex depart from sensorial inputs, however, the shape of receptive fields becomes highly complex, and the connections with sensorial input weaker. The receptive field concepts by themselves are not enough to account for neural contents, and need to be integrated with other coding concepts, as will be discussed in Sect. 4.3.

The main factor making receptive fields complex and not as easy to identify as their definition would suggest, was already present in the first studies done by Hartline on the *Limulus*: the effect of lateral interactions (Hartline et al. 1961). In the case of the *Limulus* each omnatidum has only inhibitory connections with its immediate neighbors, and still the resulting effects were not straightforward, for the recurrent property of this interaction. In the cortex, lateral interactions become dominant for two main reasons. First, there is an overlapping mechanism of inhibitory and excitatory connections as well, and second, lateral connections from a cell extend over a long range, reaching for example, in V1, up to 7 mm (Gilbert and Wiesel 1983; Stettler et al. 2002). Lateral interactions seem to play a fundamental role in the computational properties of the cortex, in a way that is yet far from being well understood (Sirosh et al. 1996; Cerreira-Perpiñán and Goodhill 2004; Hunt et al. 2011). Lateral interactions are constitutive of the mathematics upon which the models presented in this book are based (see Sect. 4.2.1). Due to lateral interactions the shape of receptive fields in the cortex is less influenced by the afferent pathway of thalamic connections and therefore, the relationship between neural firing and the retinal stimulus might be highly complex. In practice, the recurrent mechanism of lateral interactions, replicated over multiple layers of processing, makes it almost impossible to derive mathematical formulations for receptive fields in visual areas beyond V1.

Chirimuuta and Gold (2009) recall that the original concept of receptive field was of a static property of neurons, and question its validity today, in the face of all the evidence on the influence on the response of neurons or columns, given by signals out of their receptive fields. They list several possible answers, from expanding the concept of receptive field taking into account most, if not all, the factors that influence the shape of the response, to changing the kind of stimuli typically used to assess receptive fields in the case of vision, or shifting into a circuital concept, where the combined influence of a population of columns is taken into account. A step in this direction, we believe, is the last key concept of the cortex that will be presented in the following subsection.

### *3.3.3 Topological Maps*

In addition to the key concept of columnar organization, Mountcastle (1957) introduced a second notion, now widespread, concerning the organization of the cerebral cortex, that of "cortical maps". The two ideas are related: since the vertical dimension is traversed by subcomponents of a unique computational unit, it is along the 2-dimensional surface of the cortex that the firing of neurons signals the occurrence of a stimulus on a spot in a sensorial area, and the topological mapping is the first fundamental correlation between sensorial space and cortical space.

Mountcastle obtained these results during his investigations of the cat's somatosensory cortex, but he speculated that what he saw before him might be a more general, possibly fundamental, architectural principle of the mammalian cortex as a whole. Shortly after, solid confirmation arrived from the above-mentioned studies of Hubel and Wiesel on the visual cortex, and the term "cortical maps" made its appearance as a reference notion for most of the studies on the cortex, and has been used ever since. In fact, cortical maps have been found in nearly all of the sensory and motor areas of the brain (Felleman and Van Essen 1991), and the difficulty in characterizing other areas as maps as well, lies in the lack of a direct meaning of the space dimensions in the cortex. A theoretical advantage in the notion of "cortical maps" is the empirical criteria for identifying a portion of the cortex, unified functionally as a specific neural circuit: the consistent responsiveness of the cells in that part to contiguous sensorial stimulation.

The appeal to the spatial correspondence with sensorial periphery has not been the only criteria for partitioning the cortex into meaningful neural aggregates. At the beginnings of the enterprise of understanding the functions of cortical areas, anatomical methods dominated. In some fortunate cases, anatomy was in fact, sufficient for the precise identification of functional maps to be done. This is the case of V1, which can easily be identified by its heavy myelination in layer 4C, using a light microscope in post-mortem material. This was known since the eighteenth century as *lineola albidior* (Gennari 1782). The anatomical approach continues to be extremely useful today supported by sophisticated methods, such as the combination of connectivity patterns and myeloarchitecture (Maunsell and Van Essen 1983; Felleman and Van Essen 1991), [2–14C]deoxyglucose tracing (Macko et al. 1982), computational morphing (Van Essen et al. 2001). But in the recent past, the identification of cortical maps by direct evidence of the coherent response of neurons to a contiguous sensorial periphery has become of primary importance, thanks to non-invasive technologies. In vision science, a probing method first introduced by Stephen Engel (1997), and widely applied and extended by Brian Wandell (1999, 2005) allows the substitution of Mountcastle's electrophysiology with neuroimaging. In this method the two concepts of cortical maps and receptive fields meet: special moving patterns are presented, that span the entire retinal area, while the subject is scanned using fMRI. Patterns are high contrast checkboard sectors in a contracting ring or spinning wedges. The expanding-contracting ring measures topological organization of maps with respect to visual eccentricity, while

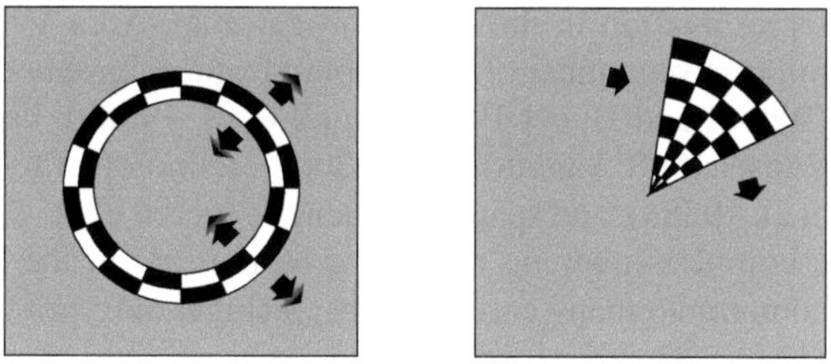

**Fig. 3.3** Sketch of the patterns used by Engel and Wandell in the search for retinotopic maps in the visual cortex. On *the left*, the expanding ring, on *the right*, the spinning wedge

wedges are used to assess topological organization with respect to polar angles. The two patterns are sketched in Fig. 3.3.

Cortical maps joined with columns and receptive fields, helped complete the picture of the cortex as a representational device. While the notion of receptive field helps in ascribing content to the firing of a single column, the same firing in the context of a cortical map acquires additional meaning by the spatial relationship the column under investigation has with the columns of the same map. This kind of organization may qualify as representation by virtue of its structural similarity property (O'Brien and Opie 2004), discussed in Sect. 3.2.2: a mapping is established between the topology of the columns, and relations between features of objects in the world. The most direct kind of mapping is of a spatial nature itself, such as the concept of retinotopy, where information represented in the map concerns the topology of the stimulus in the retina. However, as Mountcastle had warned in his early studies, cortical maps should not be interpreted as modified copies of the array of receptors in the periphery.

First of all, maps in the cortex are more often overlaps of several different sensorial features. In Mountcastle's experiments, he classified three different peripheral modalities: stimulation at the skin level, deep pressure stimulation, and that related with joint position. In the investigated cortical map, he found an overlap of different modalities projected by the same peripheral area, with neurons responding to skin stimulation for example, intermingling with those responding to deep pressure in a mosaic-like fashion. Even if we limit the analysis to a single modality, and the interpretation to the spatial representation of the stimuli, none of the features represented in a cortical map appear to be topographically simple. Maps often contain modular repetitions of small segments of receptor areas, within a global topography (Krubitzer 1995; Vanduffel et al. 2002). Moreover, inside a module where topology is preserved, metrics are often distorted, with seemingly purposeful magnifications and other transformations (Hubel and Wiesel 1974b; Van Essen et al. 1984).

The most intriguing aspect of cortical maps, however, is that the ordering in the two dimensions of the cortical sheet might represent any feature of interest in the sensorial stimuli, without any relationship to the spatial topology of the stimulus itself. This is the case of the tonotopic organization of the auditory cortex (Verkindt et al. 1995). As in the case of submodalities of the sensorial periphery, also in respect to features it must be expected that more than one feature will find simultaneous representation in the same cortical map. Area V1, for example, is one where an impressive number of overlapping features have been discovered: ocular dominance (Wiesel and Hubel 1965; Tootell et al. 1988b), orientation selectivity (Hubel and Wiesel 1968; Vanduffel et al. 2002), retinotopy (Tootell et al. 1988c), color (Tootell et al. 1988d), and spatial frequency (Tootell et al. 1988a). The suspicion is that such a complex mapping might not be unusual in the cortex, and might very well be common to many cortical maps, just that only few characteristic features have been discovered so far for other areas.

Questions regarding the extent to which the map architecture is ubiquitous as the representation strategy of the cortex, and how map contents should be interpreted, is a matter of open debate, with several opinions contending for dominance. In the early discussions on brain representation the dominant view was that topological organization might even be detrimental or incompatible with the way the cortex functions, which was assumed to be mainly associative (Kaas 1997). Today, on the contrary, the widely held opinion is that cortical maps are not incidental, but essential to the nature of brain representations. There have been several suggestions arguing that two dimensional topological maps might be the most efficient representation coding, given that neurons work by synaptic connections. Thus, placing connected neurons as close to each other as possible is an evolutionary strategy to save wiring costs, and cortical maps would thus be the resulting prevailing architecture in the brain (Swindale 2001; Chklovskii and Koulakov 2004). A good demonstration is retinotopy, that allows neurons to represent adjacent parts of the visual field, and to interact over short axonal and dendritic pathways.

Other authors have argued that cortical maps are the optimal solution, but with respect to computational properties rather than anatomical constraints. For example, from an information-theoretic point of view, ordered maps maximize the mutual information between input and output signals (Linsker 1989), or in terms of parameter space of the stimuli, cortical maps perform optimal dimension-reducing mappings (Durbin and Mitchison 1990). However, the solution of ordered maps as cortical representations is not a universal rule. It was known since the early investigations of V1 that several rodents, such as hamsters (Tiao and Blakemore 1976) and rabbits (Murphy and Berman 1979) do not have orderly orientation maps in the primary visual cortex, but do have orientation-selective neurons. The lack of orientation maps in these rodents was supposed to be related to their poor visual ability, or their small absolute V1 size. But recently, investigations on a highly visual rodent, with a large V1, the gray squirrel, confirmed the lack of orientation maps (Van Hooser et al. 2005). This result of course cannot rule out that rodents may still have a system of organization in V1 with respect to orientation, that we are not able to identify and understand.

In addition to sensorial areas, topological mapping has also been found in the agranular cortex, such as in the posterior parietal cortex, where a mosaic of columns that evoke small, specific hand-forearm movements for reaching and grasping has been observed in monkeys (Kaas et al. 2011). The kind of topological order in the homotypical cortex is clearly hard to investigate, however, it has long been supposed to include many small feature ordered maps (Kohonen and Hari 2000). Thivierge and Marcus (2007) speculate that topographic maps in the homotypical cortex could be at the basis of abstract reasoning, implementing computations such as "universally quantified one-to-one mappings", hard to simulate with artificial neural models.

The notion of cortical maps may seem related to a dated picture of the brain as a collection of autonomous modules, that was common in the early years of cognitive science (Fodor 1983), the modularity of the mind, however, tempting as it may be at a psychological level, does not fit with the organization of the brain, and more specifically of the cortex. A module was defined by Fodor with a set of nine properties, most of which clash dramatically with brain evidence, in particular their innate determination, inaccessibility from other modules, and encapsulation. Contrary to innate determination, developmental neurobiology has provided a substantial amount of evidence of a deep interaction between genetic factors and the experience of individuals in the formation of mature cortical connectivity (Blumberg et al. 2010; Braddick et al. 2011; Rubenstein and Rakic 2013a,b). Taking the case of vision, which was paradigmatic for Fodor, inaccessibility and encapsulation have been largely disproved by neuroscientific evidence: at almost all levels the visual system receives top-down projections from the cortex, and interacts with other perceptual and motor systems (Churchland et al. 1994; Callaway 2005; Paradiso et al. 2005; Niell and Stryker 2010). The existence of a specific module for language, the cognitive faculty that is the object of this book, influentially advocated by Chomsky (1986) and Hauser et al. (2002), has also been glaringly disconfirmed by empirical brain evidence (Stowe et al. 2004; Osterhout et al. 2007; Pulvermüller 2010). Mind modularity has been recently reproposed with substantial differences from the view of Fodor (Cosmides and Tooby 1997; Carruthers 2006). On one hand some of the strongest and most critical requirements of Fodorian modules are abandoned, with the most significant missing item being encapsulation. On the other hand, the revised modularity thesis is much stronger than that of Fodor, in that it includes *every* higher order cognitive function. For Fodor modularity was limited to the perceptual systems, but not the way central cognition was organized. Under the new view modularity is *massive*, and the repertoire of specialized modules include, for example, one by which we reason about physical phenomena, one for doing formal logic, and one for behaving fairly. As pointed out by Prinz (2006b), it is problematic to defend properties such as inaccessibility in such a modular system. And, yet again, neuroscientific evidence of modularity at the neural networks level, has little to do with the partitioning of cognitive functions inside massive modularity (Bullmore and Sporns 2009; Hagmann et al. 2010).

Despite what we have discussed so far, it is almost impossible to approach a study of cognition without a criteria for creating a division of tasks and a hierarchy of their internal components. This division is equally important for theoretical analysis and for building computational models. The notion of cortical maps offers the best biologically legitimate partitioning criteria for the cortex (Plebe 2008). This is a weak notion of "module", that has the unique advantage of respecting a real specialization of a bounded portion of the cortex. Computational models are fully justified in necessarily resorting to a modular structure for feasibility, if modules are constrained to correspond to cortical maps, preserving the same hierarchy and basic connections of the cortex.

## 3.4 Semantic Processing Pathways

In addition to the possible ways of identifying meaningful components inside the whole cortex, reviewed in the past section, there is a global criterion according to which the cortex can be divided into two broad main processing pathways: dorsal and ventral. It has been postulated that a range of higher-order cognitive functions are carried out by a division of labor between these two partially segregated parallel processing streams. The most interesting aspect of this theory for our purposes is that, while the dorsal stream is difficult to characterize, leading to a number of controversial hypotheses, a general consensus seems to exist on what the functional specialization of the ventral stream is: semantics.

According to this perspective, the common purpose of the ventral stream is that of extracting meaning from what is perceived in several modalities. This section will begin by discussing details on the ventral visual stream, historically the first to be conceptualized as having a semantic specialization. Then other domains for which the same concept has been extended will be reviewed, such as the auditory system, language, and attention.

### *3.4.1 The Hierarchy of Cortical Maps in the Ventral Visual Paths*

The first experiments to investigate the existence of two visual systems were carried out by Schneider (1967) with golden hamsters, finding segregation in the processing of patterns and space. This result was soon confirmed by Trevarthen and Sperry (1968), who, working on split-brain monkeys, differentiated between a system for object vision and an ambient system for guiding behavior and locomotion. A full conceptualization of the dual stream model is due to Ungerleider and Mishkin (1982), who proposed the fortunate "what" and "where" dichotomy, in that the ventral pathway is specialized for object perception, whereas the dorsal pathway is specialized for spatial perception. They supported their thesis with strong evidence from careful experiments with lesioned monkeys. When engaged in a pattern-discrimination task severe impairment was induced by ventral but not dorsal lesions, while a landmark task was impaired in the case of damage to the dorsal path, and not in the ventral path.

A few years later, a different theory for independent paths of processing in vision emerged, Hubel and Livingstone (1987) and Livingstone and Hubel (1987) proposed four streams, beginning in the retina, with the division into magno- and parvo-ganglion cells, that cross LGN and V1, and then proceeding up to higher areas. The four paths serve the separate processing of form, color, motion and stereo information. Differently from the dorsal-ventral partition, this hypothesis, despite important evidence, has been highly controversial. The four pathways do not have the same sharp functional separation found in the dorsal-ventral case, as there is significant interaction among them (Nealey and Maunsell 1994; Van Essen and DeYoe 1994).

The ventral-dorsal distinction has been enormously influential in helping to interpret the functional organization of the visual cortex. For what concerns the "what" and "where" dichotomy, while there has been large consensus on the "what" interpretation of the ventral path, the "where" has been less successful, and has been subject to a number of newly proposed interpretations. One of these interpretations, considers the dorsal system as coding visual information for action organization, therefore, "where" should be read as "how" (Goodale and Milner 1992). A possible reconciliation of the dorsal interpretation as space perception or action organization can be achieved by a further division, with a dorso-dorsal stream related to action and the ventro-dorsal stream playing a role in space perception (Rizzolatti and Matelli 2003).

In Fig. 3.4 the currently most accepted representations of the ventral paths in the human and macaque monkey visual systems are sketched. The lowest areas are common to both the dorsal and ventral parts, starting with V1, the primary visual cortex, the most studied part of the brain (Hubel and Wiesel 1962, 2004), and the site of several overlapping functions. The most important for early shape analysis is the organization into domains of orientation tuned neurons (Blasdel 1992; Vanduffel et al. 2002), other functions include ocularity (Wiesel and Hubel 1965), color (Landisman and Ts'o 2002), contrast and spatial frequency (Tootell et al. 1988a). Immediately anterior to V1, area V2 has a less understood role in vision, a general and shared idea is that it is responsible for shape analysis at a level of complexity and at a scale larger than that of V1 (Kobatake and Tanaka 1994). This is compatible with findings of V2 cells responding to end-lines and corners (Heider et al. 2000), and, with shared stronger evidence, angles (Ito and Komatsu 2004; Anzai et al. 2007).

The narrow strip of cortex surrounding V2 anteriorly, often named C3, has been supposed to be the first separation point between the two streams, with its dorsal part having a higher incidence of directionally selective neurons but less color selective neurons (Burkhalter and Van Essen 1986), and it is sometimes referred to as a group of two areas, with the addition of a ventral/dorsal suffix in the names, such as "V3v" and "V3d" (Kaas and Lyon 2001; Zeki 2003). What both parts of V3 share functionally, is selectivity in response by a consistent population of neurons to the direction of motion in the scene, and of some cells in response to stereovisual disparity, suggesting a role in the processing of motion information (Felleman et al. 1984; Gegenfurtner et al. 1997; Press et al. 2001).

After V3, dorsal and ventral streams clearly depart, and there is no symmetric correspondence between maps in the upper and lower areas. In the ventral stream,

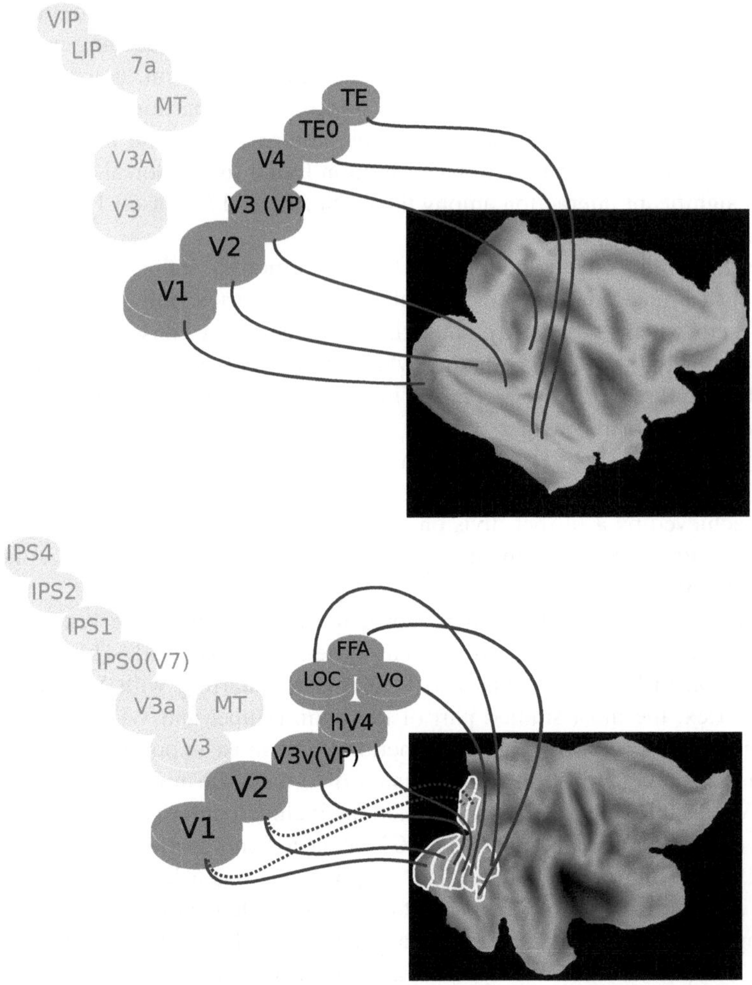


**Fig. 3.4** Scheme of the division in ventral (*darker*) and dorsal (*lighter*) visual paths in the macaque cortex (*above*), and in the human cortex (*below*) (Ventral maps are reported on flat right hemisphere cortex representations, from http://sumsdb.wustl.edu:8081/sums/, see Van Essen (2005))

an area called "V4" was first identified in monkeys by Zeki (1971), and associated to color processing. Zeki (1983a,b) made a distinction between cells that respond selectively to "wavelengths" and those responding to "colors", where the last term refers to the perceptual property of seeing a surface as constantly colored despite the large variation in the composition of the energy and wavelength of the light that is reflected from it. The conscious experience of color also seems to be supported by V4, as shown in experiments using the McCollough effect (Barnes et al. 1999; Morita et al. 2004). It is an illusory effect, in which the color stimuli are constant, but their perception can be varied gradually, by alternating two orthogonally oriented grated patterns (McCollough 1965). By inducing this effect it is possible to expose several subjects to the same color, with only some of them consciously perceiving the color. These studies have demonstrated that V4 was activated only in subjects aware of the color. Several authors have reported on the role of V4 in other types of intermediate-level visual processing, like shape (David et al. 2006) and texture (Arcizet et al. 2008) recognition.

As expected by the semantic interpretation of the visual ventral stream, moving anteriorly to V4, the computations performed in cortical areas are less affected by the local features of the scene captured by the retinas, and more sensitive to the meaning of its content: the recognition of the objects seen. It is well established that this crucial step in the monkey takes place mainly in the IT area (InferoTemporal) (Desimone et al. 1984; Tanaka et al. 1991; Kobatake and Tanaka 1994; Tanaka 1996). As is the case for V1 and V2, it has been supposed that a human homologous must exist, it was therefore surprising to discover, when accurate fMRI become available, that a different area in the geography of the human cortex is involved in object recognition. Malach et al. (1995) first identified this region, an area located anteriorly to Brodmann's area 19, near the lateral occipital sulcus, and called it LOC (Lateral Occipital Complex), where the term "complex" denotes the uncertainty that exists on whether this region is a single visual map or a cluster of several maps. There is converging evidence for at least two main components of LOC, one posterior called simply LO, located in the posterior inferior temporal gyrus, and an anterior part, LOa, extending ventrally into the middle fusiform gyrus (Malach et al. 2002; Denys et al. 2004).

Probably, the most important property needed in order to fulfill the requirement of being an object-recognition area, is that its cells exhibit several forms of invariance. Invariance in vision is the ability to recognize known objects despite large changes in their appearance on the sensory surface. It is one of the features of the biological vision system most difficult to understand, but it is also a challenging theoretical problem, because it is related to the philosophical issue of the format of mental representations (Cummins 1989). For example, invariance has been central in the debate on whether representations in the brain are 3D object-centered or image-based (Tarr and Bülthoff 1998; Edelman 1999).

Grill-Spector et al. (1999) investigated invariance using fMR-A (functional Magnetic Resonance Adaptation), a methodology based on the reduction of neural activity when visual areas are presented repetitively with the same visual stimulus, and in studying invariance, by gradually manipulating a single property in the presented image, and checking if the neural signals recover from adaptation. They found invariance to translation and size in anterior LOC (LOa or pFs) and not in posterior LOC (LO), and no invariance to rotation, which instead is found in Kourtzi et al. (2003) and in Vuilleumier et al. (2002) (but only in the left hemisphere). Invariance to the overall level of intensity, and the contrast between the recognized object and the background was also demonstrated in LOC Avidan et al. (2002), and related to the attention given to the object in order to segregate it from the background (Murray and He 2001). Invariance to a special class of rotation was found by Weigelt et al. (2007). The viewpoint rotation that LOC seems to be invariant to, is that where different views of an object are linked by apparent motion, thereby creating the illusion of a smooth rotational object motion.

Grill-Spector et al. (1998) addressed a special kind of invariance, with respect to "cues". It is the ability to respond selectively to objects, independently (at least in part) to the way this object is represented. In the experiment several kinds of stimuli were presented, in which objects might be identified by a variety of cues: motion, luminance, and texture. Luminance is the most obvious feature in a scene for detecting objects, thanks to variations in contrast with respect to the background. In the motion stimuli, all the values of luminance used were random noise, but by coherently moving a section of the noise pattern over the stationary background, an image of a drifting object silhouette was perceived. In the texture stimuli there was also no significant difference in luminance between objects and background, but the shape of the objects was derived by wrapping a texture around a three-dimensional object and filling the background with a flat texture. The results demonstrated that LOC responds to visual recognition, in a manner largely independent from the visual cue used to define objects. Other variations of cue are the representations of the same objects as photographs or as line drawings. Kourtzi and Kanwisher (2000), using also fMR-A, found that LOC responds invariantly with respect to this cue variation.

An even more compelling cue for LOC as a first semantic area is its lateralization, which should be a logical consequence of the fact that objects, in humans, are categorized by their names, and language processing is strongly lateralized. Vuilleumier et al. (2002) found hemispheric asymmetry concerning larger rotation invariance in left LOC, an indication of a more semantic biased representation in the left hemisphere. Stronger evidence of a linguistic connection of the left LOC comes from studies applying categorial differentiation in test objects, and adding linguistic test conditions. For example, the adaptation of a fMR signal is compared using a series of pictures of the same umbrella, or a series of different exemplars of umbrellas, or followed by an object of a different category. The left anterior LOC seems to be more invariant to exemplars of the same category than its right counterpart (Koutstaal et al. 2001). Moreover, if pictures are presented in conjunction to the sound of words, left anterior LOC is found to be more sensitive to the auditory perception of the object names as opposed to nonsense words (Simons et al. 2003). When the task is to name the recognized object, again, left LOC is more active than when the subject is just required to check the matching of two pictures in a sequence (Large et al. 2007).

Despite all this evidence, a clear semantic role for LOC is controversial. For example Ferber et al. (2005) argue that LOC subserves figure-ground segregation, something they deem as being a low-level task in the visual processing hierarchy, and Kim et al. (2009) contend that LOC is actually sensitive to classes of object shapes, rather than object categories. Orban et al. (2014) hold a different view, suggesting that LOC is the site, along the ventral stream, where the transition from visual features to real-world entity representations takes place. Our position is similar, we hold that semantics is not a clearly cut segregated process that is sharply located in the brain, it is the result of a chain of processes and interactions.

The LOC area is very likely a crucial computational step in this chain for the visual system, lying somewhere in between the processing of higher-level visual features of segmented objects, and their recognition.

Most of the investigations beyond LOC have searched for areas specialized in the recognition of specific classes of objects. The first was identified by Kanwisher et al. (1997) and named FFA (Fusiform Face Area), because of its location in the fusiform gyrus and because it is more active when viewing faces, compared to other objects (Grill-Spector 2003; Kanwisher 2003). How FFA could be specifically dedicated to faces is still controversial, it has been found for example, that if experts of cars or birds viewed stimuli from their domains of expertise, cars or birds respectively, the right FFA was significantly more active than for other common objects (Gauthier et al. 2000; Xu 2005). These results suggested an alternative hypothesis, that FFA is in fact an area for holistic visual processing automatized by expertise, and it is not surprising that the object deserving the highest expertise for our social life is the face, and its process is typically holistic (Tarr and Gauthier 2000; Gauthier and Tarr 2002). The expert hypothesis is not without controversy, particularly concerning the appropriate task design and analyses used to measure holistic and configural effects (Robbins and McKone 2007; Gauthier and Bukach 2007). Moreover, face recognition and identification appear to be distributed across a network of areas much wider than FFA, including the inferior occipital gyrus and the superior temporal sulcus (Haxby et al. 2001; O'Toole et al. 2005; Nestor et al. 2011).

It is in this area, nevertheless, where proof for specificity for a class of objects is more convincing. There are clues for a region in the medial temporal lobe, called PPA (Parahippocampal Place Area), that responds to "places", that is, scenes where the overall layout is important (Epstein and Kanwisher 1998), however, this region seems to be more generally responsive to high spatial frequencies (Rajimehr et al. 2011). Yet another area, in the lateral occipitotemporal cortex on the lower lip of the posterior temporal sulcus, called EBA (Extrastriate Body Area), seems to respond to images of parts of the human body (Downing et al. 2001), but is also involved in perception of emotions and the understanding of actions (Peelen and Downing 2007).

### *3.4.2 Other Ventral Streams*

After having been discovered in the visual system, the division between a ventral stream, related with the "what", and a less definable dorsal stream, was reportedly found in several other cognitive processes, at the point that it is suspected to be a general feature of cortical organization (Cloutman 2013).

The second system for which a double "what"/"where" path has been proposed is the auditory system by Romanski et al. (1999), based on evidence in monkeys. Interestingly, the same controversy on the dorsal visual path immediately arose, with (Belin and Zatorre 2000) contending that it might more properly concern

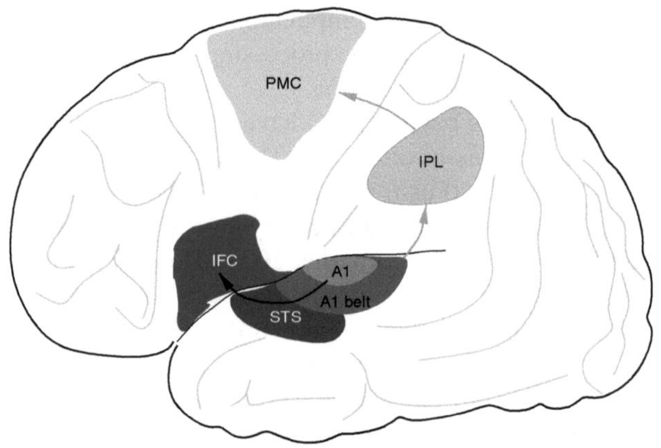


**Fig. 3.5** Scheme of the division in ventral (*darker*) and dorsal (*lighter*) auditory paths in the human cortex (below)

the "how", rather than the "where", but no critics have come forth instead, for the "what" classification. A word of caution is necessary in this case, since it is not at all obvious how one should define a content meaning for a sound. Kubovy and Valkenburg (2000) argue for the existence of *auditory objects*, defined as the entities that generate a sound, with its categorization being the purpose of the ventral auditory stream. However, as pointed out by Griffiths and Warren (2004), this definition would miss the event information of the sound, which can be its actual content. For example in hearing an uttered vowel, it seems obvious that the interesting content is the vowel category, not just the speaker, or the speaker's vocal tract generating that sound.

Despite these difficulties, the existence of a ventral stream that is apt to process the meaning of a sound, in a loose definition, has been reinforced and confirmed in humans (Arnott et al. 2004; Rauschecker and Scott 2009). A picture of the two pathways as currently conceptualized in the human auditory system is given in Fig. 3.5.

The primary auditory cortex, usually referred to as A1 in analogy to V1, is a site of several overlapped processes, just like its visual counterpart, but it is much less understood. There is certainly an organization with respect to spectro-temporal features of sounds (Miller et al. 2002b; Winer et al. 2005), mixed with acoustic levels and spectral integration, with neurons responding to stationary components of the sound, or rapid transients (Atzori et al. 2001). A1 is surrounded by secondary auditory cortices, the belt areas, which are bordered laterally by a parabelt region. Its rostral region is involved in the ventral stream, and projects to the ventral part of the superior temporal gyrus, area STS (*Superior Temporal Sulcus*), which is known to be involved in phonological processes (Belin et al. 2004; Liebenthal et al. 2005). The path proceeds further to the anterior temporal lobe, terminating in multiple frontal lobe regions including the frontal pole, and ventral prefrontal areas.

A more audacious extension of the ventral/dorsal concept has been postulated for language itself. According to Hickok and Poeppel (2007), the first proposers, the dorsal stream, traveling from the superior temporal gyrus toward inferior parietal and posterior frontal lobe regions, is responsible for repetition of speech. On the other hand, the ventral stream, traveling from the superior temporal gyrus laterally to the middle and inferior temporal cortices, has the purpose of giving meaning to speech. This hypothesis has met with several confirmations, DeWitt and Rauschecker (2012) found that the ventral stream is essentially semantic, being engaged in both the invariant representation of phonetic forms and in the integration of phonemes into words. Saur et al. (2008) combined fMRI with DTI (*Diffusion Tensor Imaging*) to accurately identify the anatomical pathways during a task of sublexical speech repetition, and of language comprehension. The latter is clearly subserved by a ventral pathway. Again, the most controversial, is the function of the dorsal stream, with alternative hypotheses ranging from syntax analysis (Friederici 2012) to its being linked to the general processing of time-dependent components (Bornkessel-Schlesewsky and Schlesewsky 2013). Much less disputed is the attribution of a semantic function of the ventral stream. Of course, while in sensory domains the meaning of the perceived signal can be better defined, in the case of language, meaning is tremendously wide-ranging, and cannot be confined to a specific anatomical processing path. Language comprehension, in its full sense, involves a multitude of cognitive processes and cortical regions, that this book aims to contribute in clarifying, at least in part. Nevertheless, it is important to acknowledge that at least a starting point in the process of assigning meaning to speech relies on one specific ventral pathway in the cortex.

There are also other non sensorial domains, for which a dorsal/ventral division have been proposed. Umarova et al. (2010) for example, investigated attention, using fMRI and DTI, and found that the ventral stream, traveling between the insula and putamen, parallel to the sylvian fissure, is specialized in recognizing the object of attention, while the dorsal stream, linking the temporoparietal cortex with the frontal eye field and the inferior frontal gyrus, is responsible for attention orientation.

## References


Adrian, E. D., & Matthews, R. (1927a). The action of light on the eye: Part I. The discharge of impulses in the optic nerve and its relation to the electric changes in the retina. *Journal of Physiology, 63*, 378–414.

Adrian, E. D., & Matthews, R. (1927b). The action of light on the eye: Part II. The processes involved in retinal excitation. *Journal of Physiology, 64*, 279–301.

Anzai, A., Peng, X., & Van Essen, D. C. (2007). Neurons in monkey visual area V2 encode combinations of orientations. *Nature Neuroscience, 10*, 1313–1321.

Arcizet, F., Jouffrais, C., & Girard, P. (2008). Natural textures classification in area v4 of the macaque monkey. *Experimental Brain Research, 189*, 109–120.

Aristotele (350BCE). *On memory and reminiscence* (W. D. Ross, Trans., 1930). Oxford: Clarendon Press.

Aristotle (335–323BCE). *De interpretatione* (J. L. Ackrill, Trans., 1975). Oxford: Clarendon Press.

Arnauld, A. (1683). Des vraies et des fausses idées. Adolphe Delahays, Paris, published in Oevres Philosophiques, 1843.
Arnott, S. R., Binns, M. A., Grady, C. L., & Alain, C. (2004). Assessing the auditory dual-pathway model in humans. *NeuroImage, 22*, 401–408.
Atzori, M., Lei, S., Evans, D. I. P., Kanold, P. O., Phillips-Tansey, E., McIntyre, O., McBain, C. J. (2001). Differential synaptic processing separates stationary from transient inputs to the auditory cortex. *Neural Networks, 4*, 1230–1237.
Avidan, G., Harel, M., Hendler, T., Ben-Bashat, D., Zohary, E., & Malach, R. (2002). Contrast sensitivity in human visual areas and its relationship to object recognition. *Journal of Neurophysiology, 87*, 3102–3116.
Axmacher, N., Mormann, F., Fernández, G., Elger, C. E., & Fell, J. (2006). Memory formation by neuronal synchronization. *Brain Research Reviews, 52*, 170–182.
Ayer, A. (1940). *The foundations of empirical knowledge*. London: Macmillan.
Barnes, J., Howard, R., Senior, C., Brammer, M., Bullmore, E., Simmons, A., & David, A. (1999). The functional anatomy of the mccollough contingent colour after-effect. *NeuroReport, 10*, 195–199.
Bechtel, W. (1998). Representations and cognitive explanations: Assessing the dynamicist's challenge in cognitive science. *Cognition, 22*, 295–318.
Bechtel, W. (2014). Investigating neural representations: the tale of place cells. *Synthese*, 1–35. doi:10.1007/s11229-014-0480-8
Bechtel, W., & Richardson, R. C. (1993). *Discovering complexity: Decomposition and localization as scientific research strategies*. Princeton: Princeton University Press.
Behrens, T. E., Hunt, L. T., Woolrich, M. W., & Rushworth, M. F. (2008). Associative learning of social value. *Nature, 13*, 245–249.
Belin, P., & Zatorre, R. J. (2000). 'what', 'where' and 'how' in auditory cortex. *Nature Neuroscience, 3*, 965–966.
Belin, P., Fecteau, S., & Bédard, C. (2004). Thinking the voice: Neural correlates of voice perception. *Trends in Cognitive Sciences, 8*, 129–135.
Blasdel, G. G. (1992). Orientation selectivity, preference, and continuity in monkey striate cortex. *Journal of Neuroscience, 12*, 3139–3161.
Blouw, P., Solodkin, E., Thagard, P., & Eliasmith, C. (2015). Concepts as semantic pointers: A framework and computational model. *Cognitive Science*, 1–35. doi:10.1111/cogs.12265
Blumberg, M. S., Freeman, J. H., & Robinson, S. (Eds.). (2010). *Oxford handbook of developmental behavioral neuroscience*. Oxford: Oxford University Press.
Bornkessel-Schlesewsky, I., & Schlesewsky, M. (2013). Reconciling time, space and function: A new dorsalventral stream model of sentence comprehension. *Brain and Language, 125*, 60–76.
Braddick, O., Atkinson, J., & Innocenti, G. M. (Eds.). (2011). *The developing brain: From developmental biology to behavioral disorders and their remediation*. Cambridge: Cambridge University Press.
Bullmore, E., & Sporns, O. (2009). Complex brain networks: Graph theoretical analysis of structural and functional systems. *Nature Reviews Neuroscience, 10*, 186–198.
Burke, C. J., Tobler, P. N., Baddeley, M., & Schultz, W. (2010). Neural mechanisms of observational learning. *Proceedings of the Natural Academy of Science USA, 32*, 14431–14436.
Burkhalter, A., Van Essen, D. C. (1986). Processing of color, form and disparity information in visual areas VP and V2 of ventral extrastriate cortex in the macaque monkey. *Journal of Neuroscience, 6*, 2327–2351.
Caggiano, V., Fogassi, L., Rizzolatti, G., Casile, A., Giese, M. A., & Thier, P. (2012). Mirror neurons encode the subjective value of an observed action. *Proceedings of the Natural Academy of Science USA, 29*, 11848–11853.
Callaway, E. M. (2005). Structure and function of parallel pathways in the primate early visual system. *Journal of Physiology, 566*, 13–19.
Calvo-Merino, B., Grezes, J., Glaser, D. E., Passingham, R., & Haggard, P. (2006). Seeing or doing? Influence of visual and motor familiarity in action observation. *Current Biology, 16*, 1905–1910.

Carruthers, P. (2006). *The architecture of the mind*. Oxford: Oxford University Press.
Cerreira-Perpiñán, M., Goodhill, G. J. (2004). Influence of lateral connections on the structure of cortical maps. *Journal of Neurophysiology, 92*, 2947–2959.
Chemero, A. (2009). *Radical embodied cognitive science*. Cambridge: MIT.
Chirimuuta, M., Gold, I. (2009). The embedded neuron, the enactive field? In J. Bickle (Ed.), *Handbook of philosophy and neuroscience*. Oxford: Oxford University Press.
Chklovskii, D. B., Koulakov, A. A. (2004). Maps in the brain: What can we learn from them? *Annual Review of Neuroscience, 27*, 369–392.
Chomsky, N. (1986). *Knowledge of language: Its nature, origins and use*. New York: Praeger.
Church, A. (1941). *The calculi of lambda conversion*. Princeton: Princeton University Press.
Churchland, P. S. (2002). *Brain-wise. Studies in neurophilosophy*. Cambridge: MIT.
Churchland, P. M. (2010). Concept formation via Hebbian learning: The special case of prototypical causal sequences. In P. Machamer & G. Wolters (Eds.), *Interpretation – Ways of thinking about the sciences and the arts* (pp. 203–219). Pittsburgh: Pittsburgh University Press.
Churchland, P. S., Ramachandran, V., & Sejnowski, T. (1994). A critique of pure vision. In C. Koch & J. Davis (Eds.), *Large-scale neuronal theories of the brain*. Cambridge: MIT.
Cloutman, L. L. (2013). Interaction between dorsal and ventral processing streams: Where, when and how? *Brain and Language, 127*, 251–263.
Colombo, M. (2014a). Explaining social norm compliance. A plea for neural representations. *Phenomenology and the Cognitive Sciences, 13*, 217–238.
Colombo, M. (2014b). Neural representationalism, the hard problem of content and vitiated verdicts. A reply to Hutto & Myin (2013). *Phenomenology and the Cognitive Sciences, 13*, 257–274.
Cooper, R. P., Cook, R., Dickinson, A., & Heyes, C. M. (2013). Associative (not Hebbian) learning and the mirror neuron system. *Neuroscience Letters, 540*, 28–36.
Cosmides, L., & Tooby, J. (1997). The modular nature of human intelligence. In A. Scheibel & J. W. Schopf (Eds.), *The origin and evolution of intelligence*. Oxford: Oxford University Press.
Craver, C. F. (2007). *Explaining the brain: Mechanisms and the mosaic unity of neuroscience*. Oxford: Oxford University Press.
Craver, C. (2010). Prosthetic models. *Philosophy of Science, 77*, 840–851.
Cummins, R. (1989). *Meaning and mental representation*. Cambridge: MIT.
Daugman, J. G. (1980). Two-dimensional spectral analysis of cortical receptive field profiles. *Vision Research, 20*, 847–856.
Daugman, J. G. (1985). Uncertainty relation for resolution in space, spatial frequency, and orientation optimized by two-dimensional visual cortical filters. *Journal of the Optical Society of America, 2*, 1160–1169.
David, S. V., Hayden, B. Y., & Gallant, J. L. (2006). Spectral receptive field properties explain shape selectivity in area V4. *Journal of Neurophysiology, 96*, 3492–3505.
Davidson, D. (1987). Knowing one's own mind. *Proceedings and Addresses of the American Philosophical Association, 60*, 441–458.
de Charms, R. C., & Zador, A. (2000). Neural representation and the cortical code. *Annual Review of Neuroscience, 23*, 613–647.
de Klerk, C. C., Johnson, M. H., Heyes, C. M., & Southgate, V. (2014). Baby steps: Investigating the development of perceptual-motor couplings in infancy. *Developmental Science, 8*, 1–11.
Denys, K., Vanduffel, W., Fize, D., Nelissen, K., Peuskens, H., Van Essen, D., & Orban, G. A. (2004). The processing of visual shape in the cerebral cortex of human and nonhuman primates: A functional magnetic resonance imaging study. *Journal of Neuroscience, 24*, 2551–2565.
Desimone, R., Albright, T. D., Gross, C. D., & Bruce, C. (1984). Stimulus-selective properties of inferior temporal neurons in the macaque. *Journal of Neuroscience, 4*, 2051–2062.
Devitt, M. (1981). *Designation*. Cambridge: MIT.
DeWitt, I., & Rauschecker, J. P. (2012). Phoneme and word recognition in the auditory ventral stream. *Proceedings of the Natural Academy of Science USA, 109*, E505–E514.
Douglas, R. J., Martin, K. A., & Whitteridge, D. (1989). A canonical microcircuit for neocortex. *Neural Computation, 1*, 480–488.

Downing, P. E., Jiang, Y., Shuman, M., & Kanwisher, N. (2001). A cortical area selective for visual processing of the human body. *Science, 293*, 2470–2473.
Dretske, F. I. (1981). *Knowledge and the flow of information*. Cambridge: MIT.
Dretske, F. I. (1986). Misrepresentation. In R. Bogdan (Ed.), *Belief: Form, content and function*. Oxford: Oxford University Press.
Dreyfus, H. (2002). Intelligence without representation – Merleau-ponty's critique of mental representation. The relevance of phenomenology to scientific explanation. *Phenomenology and the Cognitive Sciences, 1*, 367–383.
Dudai, Y. (2007). Post-activation state: A critical rite of passage of memories. In Bontempi, B., Silva, A., & Christen, Y. (Eds.), *Memories: Molecules and circuits* (pp. 69–82). Berlin: Springer.
Durbin, R., & Mitchison, G. (1990). A dimension reduction framework for understanding cortical maps. *Nature, 343*, 644–647.
Dylla, K. V., Galili, D. S., Szyszka, P., & Lüdke, A. (2013). Trace conditioning in insects – Keep the trace! *Frontiers in Psychology, 4*, 67.
Ebbinghaus, H. (1885). *Memory: A contribution to experimental psychology*. New York: Dover.
Edelman, S. (1999). *Representation and recognition in vision*. Cambridge: MIT.
Eliasmith, C. (2013). *How to build a brain: A neural architecture for biological cognition*. Oxford: Oxford University Press.
Eliasmith, C., & Anderson, C. H. (2003). *Neural engineering computation, representation, and dynamics in neurobiological systems*. Cambridge: MIT.
Engel, S. A., Glover, G. H., & Wandell, B. A. (1997). Retinotopic organization in human visual cortex and the spatial precision of functional MRI. *Cerebral Cortex, 7*, 181–192.
Epstein, R., & Kanwisher, N. (1998). A cortical representation of the local visual environment. *Nature, 9*, 598–601.
Feldman, D. E. (2012). The spike-timing dependence of plasticity. *Neuron, 75*, 556–571.
Felleman, D. J., & Van Essen, D. C. (1991). Distributed hierarchical processing in the primate cerebral cortex. *Cerebral Cortex, 1*, 1–47.
Felleman, D. J., Carman, G., & Van Essen, D. C. (1984). Distributed hierarchical processing in the primate cerebral cortex. *Investigative Ophthalmology and Visual Science, 25*, 278.
Ferber, S., Humphrey, G. K., & Vilis, T. (2005). Segregation and persistence of form in the lateral occipital complex. *Neuropsychologia, 43*, 41–45.
Fodor, J. (1975). *The language of thought*. Cambridge: Harvard University Press.
Fodor, J. (1983). *Modularity of mind: And essay on faculty psychology*. Cambridge: MIT.
Fodor, J. (1987). *Psychosemantics: The problem of meaning in the philosophy of mind*. Cambridge: MIT.
Fodor, J. (1990). *A theory of content and other essays*. Cambridge: Cambridge University Press.
Fodor, J. (2008). Against darwinism. *Mind & Language, 23*, 1–24.
Fodor, J., & Pylyshyn, Z. W. (2015). *Minds without meanings: An essay on the content of concepts*. Cambridge: MIT.
Fresco, N. (2014). *Physical computation and cognitive science*. Berlin: Springer.
Friederici, A. (2012). The cortical language circuit: From auditory perception to sentence comprehension acquisition, comprehension, and production. *Trends in Cognitive Sciences, 16*, 262–268.
Fries, P. (2005). A mechanism for cognitive dynamics: Neuronal communication through neuronal coherence. *Trends in Cognitive Sciences, 9*, 475–480.
Gabor, D. (1946). Theory of communication. *Journal IEE, 93*, 429–459.
Gallistel, C. R. (2000). The replacement of general-purpose learning models with adaptively specialized learning modules. In M. S. Gazzaniga (Ed.), *The new cognitive neurosciences* (2nd ed., pp. 1179–1191). Cambridge: MIT.
Gallistel, C. R. (2010). Learning organs. In J. Bricmont & J. Franck (Eds.), *Chomsky notebook* (pp. 573–586). Cambridge: Cambridge University Press.
Gallistel, C. R., & Balsam, P. D. (2014). Time to rethink the neural mechanisms of learning and memory. *Neurobiology of Learning and Memory, 108*, 136–144.

Gauthier, I., & Bukach, C. (2007). Should we reject the expertise hypothesis? *Cognition, 103*, 322–330.

Gauthier, I., & Tarr, M. (2002). Unraveling mechanisms for expert object recognition: Bridging brain activity and behavior. *Journal of Experimental Psychology Human Perception and Performance, 28*, 431–446.

Gauthier, I., Skudlarski, P., Gore, J. C., & Anderson, A. W. (2000). Expertise for cars and birds recruits brain areas involved in face recognition. *Nature Neuroscience, 3*, 191–197.

Gegenfurtner, K. R., Kiper, D. C., & Levitt, J. B. (1997). Functional properties of neurons in macaque area V3. *Journal of Neurophysiology, 77*, 1906–1923.

Gelder, T. v. (1995). What might cognition be, if not computation? *Journal of Philosophy, 91*, 345–381.

Gennari, F. (1782). *De peculiari structura cerebri, nonnulisque ejus morbis*. Parma: Ex regio typographeo.

George, D., & Hawkins, J. (2009). Towards a mathematical theory of cortical micro-circuits. *PLoS Computational Biology, 5*, e1000532.

Gilbert, C. D., & Wiesel, T. N. (1983). Clustered intrinsic connections in cat visual cortex. *Journal of Neuroscience, 3*, 1116–1133.

Glenberg, A. M., & Gallese, V. (2012). Action-based language: A theory of language acquisition, comprehension, and production. *Cognition, 48*, 905–922.

Goodale, M. A., & Milner, A. D. (1992). Separate visual pathways for perception and action. *Trends in Neuroscience, 15*, 20–25.

Goodman, N. (1976). *Languages of art: An approach to a theory of symbols*. Indianapolis: Hackett.

Gould, S. J., & Lewontin, R. C. (1979). The spandrels of San Marco and the Panglossian paradigm: A critique of the adaptationist programme. *Philosophical Transactions of the Royal Society of London Series B, Biological Sciences, 205*, 581–598.

Griffiths, T. D., & Warren, J. D. (2004). What is an auditory object? *Neuron, 43*, 237–249.

Griffiths, T. L., Chater, N., Kemp, C., Perfors, A., & Tenenbaum, J. B. (2010). Probabilistic models of cognition: Exploring representations and inductive biases. *Trends in Cognitive Sciences 14*, 357–364.

Grill-Spector, K. (2003). The functional organization of the ventral visual pathway and its relationship to object recognition. In N. Kanwisher & J. Duncan (Eds.), *Attention and performance XX. Functional brain imaging of visual cognition*. Oxford: Oxford University Press.

Grill-Spector, K., Kushnir, T., Edelman, S., Itzchak, Y., & Malach, R. (1998). Cue-invariant activation in object-related areas in the human occipital lobe. *Neuron, 21*, 191–202.

Grill-Spector, K., Kushnir, T., Edelman, S., Avidan-Carmel, G., Itzchak, Y., & Malach, R. (1999). Differential processing of objects under various viewing conditions in the human lateral occipital complex. *Neuron, 24*, 187–203.

Grush, R. (1997). The architecture of representation. *Philosophical Psychology, 10*, 5–23.

Grush, R., & Mandik, P. (2002). Representational parts. *Phenomenology and the Cognitive Sciences, 1*, 389–394.

Hagmann, P., Cammoun, L., Gigandet, X., Gerhard, S., Grant, P. E., Wedeen, V., Meuli, R., Thiran, J. P., Honey, C. J., & Sporns, O. (2010). MR connectomics: Principles and challenges. *Journal of Neuroscience Methods, 194*, 34–45.

Hall, G. (1991). *Perceptual and associative learning*. Oxford: Oxford University Press.

Hartline, H. K. (1938). The response of single optic nerve fibers of the vertebrate eye to illumination of the retina. *American Journal of Physiology, 121*, 400–415.

Hartline, H. K. (1967). Visual receptors and retinal interaction. *Science, 164*, 270–278.

Hartline, H. K., Ratliff, F., & Miller, W. H. (1961). Inhibitory interaction in the retina and its significance in vision. In E. Florey (Ed.), *Nervous inhibition*. New York: Pergamon Press.

Haselager, P., de Groot, A., & van Rappard, H. (2003). Representationalism vs. anti-representationalism: A debate for the sake of appearance. *Philosophical Psychology, 16*, 5–23.

Hauser, M. D., Chomsky, N., & Fitch, W. T. (2002). The faculty of language: What is it, who has it, and how did it evolve? *Science, 298*, 1569–1579.

Haxby, J. V., Gobbini, M. I., Furey, M. L., Ishai, A., Schouten, J. L., & Pietrini, P. (2001). Distributed and overlapping representations of objects and faces in ventral temporal cortex. *Science, 293*, 2425–2430.

Hebb, D. O. (1949). *The organization of behavior*. New York: Wiley.

Heider, B., Meskenaite, V., & Peterhans, E. (2000). Anatomy and physiology of a neural mechanism defining depth order and contrast polarity at illusory contours. *European Journal of Neuroscience, 12*, 4117–4130.

Hempel, C. G. (1965). *Aspects of scientific explanation and other essays in the philosophy of science*. New York: Free Press.

Heyes, C. (2010). Where do mirror neurons come from? *Neuroscience & Biobehavioral Reviews, 34*, 575–583.

Heyes, C. (2012). Simple minds: A qualified defence of associative learning. *Philosophical Transactions of the Royal Society B, 367*, 2695–2703.

Hickok, G. (2009). Eight problems for the mirror neuron theory of action understanding in monkeys and humans. *Journal of Cognitive Neuroscience, 21*, 1229–1243.

Hickok, G. (2010). The role of mirror neurons in speech perception and action word semantics. *Language and Cognitive Processes, 25*, 749–776.

Hickok, G., & Poeppel, D. (2007). The cortical organization of speech processing. *Nature Reviews Neuroscience, 8*, 393–402.

Horton, J. C., & Adams, D. L. (2005). The cortical column: A structure without a function. *Philosophical Transactions of the Royal Society B, 360*, 837–862.

Hubel, D. H., & Livingstone, M. S. (1987). Segregation of form, color, and stereopsis in primate area 18. *Journal of Neuroscience, 7*, 3378–3415.

Hubel, D., & Wiesel, T. (1959). Receptive fields of single neurones in the cat's striate cortex. *Journal of Physiology, 148*, 574–591.

Hubel, D., & Wiesel, T. (1962). Receptive fields, binocular interaction, and functional architecture in the cat's visual cortex. *Journal of Physiology, 160*, 106–154.

Hubel, D., & Wiesel, T. (1968). Receptive fields and functional architecture of monkey striate cortex. *Journal of Physiology, 195*, 215–243.

Hubel, D., & Wiesel, T. (1974a). Ordered arrangement of orientation columns in monkeys lacking visual experience. *Journal of Comparative Neurology, 158*, 307–318.

Hubel, D., & Wiesel, T. (1974b). Uniformity of monkey striate cortex: A parallel relationship between field size, scatter, and magnification factor. *Journal of Comparative Neurology, 158*, 295–305.

Hubel, D. H., & Wiesel, T. N. (2004). *Brain and visual perception: The story of a 25-year collaboration*. Oxford: Oxford University Press.

Hume, D. (1739). *A treatise of human nature* (Vols. 1, 2). London: John Noon.

Hume, D. (1748). *An enquiry concerning human understanding*. London: A. Millar.

Hunt, J. J., Bosking, W. H., & Goodhill, G. J. (2011). Statistical structure of lateral connections in the primary visual cortex. *Neural Systems & Circuits, 1*, 1–12.

Hutto, D. D., & Myin, E. (2013). *Radicalizing enactivism: Basic minds without content*. Cambridge: MIT.

Hutto, D. D., & Myin, E. (2014). Neural representations not needed – No more pleas, please. *Phenomenology and the Cognitive Sciences, 13*, 241–256.

Iacoboni, M., Woods, R. P., Brass, M., Bekkering, H., Mazziotta, J. C., & Rizzolatti, G. (1999). Cortical mechanisms of human imitation. *Science, 286*, 2526–2528.

Ito, M. (1984). *The cerebellum and neural control*. New York: Raven Press.

Ito, M., & Komatsu, H. (2004). Representation of angles embedded within contour stimuli in area V2 of macaque monkeys. *Journal of Neuroscience, 24*, 3313–3324.

Johnson-Laird, P. (1983). *Mental models: Towards a cognitive science of language, inference and consciousness*. Cambridge: Cambridge University Press.

Kaas, J. H. (1997). Topographic maps are fundamental to sensory processing. *Brain Research Bulletin, 44*, 107–112.

Kaas, J. H., & Lyon, D. C. (2001). Visual cortex organization in primates: Theories of V3 and adjoining visual areas. *Progress in Brain Research, 134*, 285–295.

Kaas, J. H., Gharbawie, O. A., & Stepniewska, I. (2011). The organization and evolution of dorsal stream multisensory motor pathways in primates. *Frontiers in Nauroanatomy, 5*, 34.

Kanwisher, N. (2003). The ventral visual object pathway in humans: Evidence from fMRI. In L. Chalupa & J. Werner (Eds.), *The visual neurosciences*. Cambridge: MIT.

Kanwisher, N., McDermott, J., & Chun, M. (1997). The fusiform face area: A module in human extrastriate cortex specialized for face perception. *Journal of Neuroscience, 17*, 4302–4311.

Kaplan, D. M. (2011). Explanation and description in computational neuroscience. *Synthese, 183*, 339–373.

Kaplan, D. M., & Craver, C. F. (2011). Towards a mechanistic philosophy of neuroscience. In S. French & J. Saatsi (Eds.), *Continuum companion to the philosophy of science* (pp. 268–292). London: Continuum Press.

Keysers, C., & Perrett, D. I. (2004). Demystifying social cognition: A Hebbian perspective. *Trends in Cognitive Sciences, 8*, 501–507.

Kim, J. G., Biederman, I., Lescroart, M. D., & Hayworth, K. J. (2009). Adaptation to objects in the lateral occipital complex (loc): Shape or semantics? *Vision Research, 49*, 2297–2305.

Kleene, S. C. (1936). General recursive functions of natural numbers. *Mathematische Annalen, 112*, 727–742.

Kobatake, E., & Tanaka, K. (1994). Neuronal selectivities to complex object features in the ventral visual pathway of the macaque cerebral cortex. *Journal of Neurophysiology, 71*, 856–867.

Koch, G., Ponzo, V., Lorenzo, F. D., Caltagirone, C., & Veniero, D. (2013). Hebbian and anti-Hebbian spike-timing-dependent plasticity of human cortico-cortical connections. *Journal of Neuroscience, 33*, 9725–9733.

Kohonen, T., & Hari, R. (2000). Where the abstract feature maps of the brain might come from. *Trends in Neurosciences, 22*, 135–139.

Komosinski, M. (2000). The world of framsticks: Simulation, evolution, interaction. In *Proceedings of the 2nd international conference on virtual worlds* (pp. 214–224). Paris: Springer.

Kourtzi, Z., & Kanwisher, N. (2000). Cortical regions involved in perceiving object shape. *Journal of Neuroscience, 20*, 3310–3318.

Kourtzi, Z., Erb, M., Grodd, W., & Bülthoff, H. H. (2003). Representation of the perceived 3-d object shape in the human lateral occipital complex. *Cerebral Cortex, 13*, 911–920.

Koutstaal, W., Wagner, A. D., Rotte, M., Maril, A., Buckner, R., & Schacter, D. L. (2001). Perceptual specificity in visual object priming: Functional magnetic resonance imaging evidence for a laterality difference in fusiform cortex. *Neuropsychologia, 2*, 184–199.

Krubitzer, L. (1995). The organization of neocortex in mammals: Are species differences really so different? *Trends in Neuroscience, 8*, 408–417.

Kubovy, M., & Valkenburg, D. V. (2000). Auditory and visual objects. *Nature Neuroscience, 3*, 191–197.

Kuffler, S. W. (1953). Discharge patterns and functional organization of mammalian retina. *Journal of Neurophysiology, 16*, 37–68.

Landisman, C. E., & Ts'o, D. Y. (2002). Color processing in macaque striate cortex: Relationships to ocular dominance, cytochrome oxidase, and orientation. *Journal of Neurophysiology, 87*, 3126–3137.

Large, M. E., Aldcroft, A., & Vilis, T. (2007). Task-related laterality effects in the lateral occipital complex. *Brain, 1128*, 130–138.

LeDoux, J. E. (2000). Emotion circuits in the brain. *Annual Review of Neuroscience, 23*, 155–184.

Liebenthal, E., Binder, J. R., Spitzer, S. M., Possing, E. T., & Medler, D. A. (2005). Neural substrates of phonemic perception. *Cerebral Cortex, 15*, 1621–1631.

Lingnau, A., Gesierich, B., & Caramazza, A. (2009). Asymmetric fMRI adaptation reveals no evidence for mirror neurons in humans. *Proceedings of the Natural Academy of Science USA, 106*, 9925–9930.

Linsker, R. (1989). How to generate ordered maps by maximizing the mutual information between input and output signals. *Neural Computation, 1*, 402–411.

Liu, H., & Wen, X. (2013). On formalizing causation based on constant conjunction theory. *The Review of Symbolic Logic, 6*, 160–181.

Livingstone, M. S., & Hubel, D. H. (1987). Psychophysical evidence for separate channels for the perception of form, color, movement, and depth. *Journal of Neuroscience, 7*, 3416–3468.

Locke, J. (1690). *An essay concerning human understanding*. Cleveland: Meridian Books.

Lorente de Nó, R. (1938). Architectonics and structure of the cerebral cortex. In J. Fulton (Ed.), *Physiology of the nervous system* (pp. 291–330). Oxford: Oxford University Press.

Maçarico da Costa, N., Martin, K. A. C. (2010). Whose cortical column would that be? *Frontiers in Nauroanatomy, 4*, 16.

Machamer, P., Darden, L., & Craver, C. F. (2000). Thinking about mechanisms. *Philosophy of Science, 67*, 1–84.

Macko, K., Jarvis, C., Kennedy, C., Miyaoka, M., Shinohara, M., Sololoff, L., & Mishkin, M. (1982). Mapping the primate visual system with [2–14c]deoxyglucose. *Science, 218*, 394–397.

Malach, R., Reppas, J. B., Benson, R. R., Kwong, K. K., Jiang, H., Kennedy, W. A., Ledden, P. J., Brady, T. J., Rosen, B. R., & Tootell, R. B. (1995). Object-related activity revealed by functional magnetic resonance imaging in human occipital cortex. *Proceedings of the Natural Academy of Science USA, 92*, 8135–8139.

Malach, R., Levy, I., & Hasson, U. (2002). The topography of high-order human object areas. *Trends in Cognitive Sciences, 6*, 176–184.

Malebranche, N. (1675). *De la recherche de la vérité*. Amsterdam: Henry Desbordes.

Mandik, P. (2003). Varieties of representation in evolved and embodied neural networks. *Biology and Philosophy, 18*, 95–130.

Marr, D. (1970). A theory for cerebral neocortex. *Proceedings of the Royal Society of London B, 176*, 161–234.

Maunsell, J., Van Essen, D. C. (1983). The connections of the middle temporal visual area (mt) and their relation ship to a cortical hierarchy in the macaque monkey. *Journal of Neuroscience, 3*, 2563–2586.

McCollough, C. (1965). Color adaptation of edge-detectors in the human visual system. *Science, 149*, 1115–1116.

Miłkowski, M. (2015a). The hard problem of content: Solved (long ago). *Studies in Logic, Grammar and Rhetoric, 41*, 73–88.

Miłkowski, M. (2015b). Satisfaction conditions in anticipatory mechanisms. *Biology and Philosophy, 30*, 709–728.

Miller, L. M., Escabí, M. A., Read, H. L., & Schreiner, C. E. (2002b). Spectrotemporal receptive fields in the lemniscal auditory thalamus and cortex. *Journal of Neurophysiology, 87*, 516–527.

Millikan, R. G. (1984). *Language, thought, and other biological categories: New foundations for realism*. Cambridge: MIT.

Millikan, R. G. (2000). *On Clar and confused ideas: An essay about substance concepts*. Cambridge: Cambridge University Press.

Morita, T., Kochiyama, T., Okada, T., Yonekura, Y., Matsumura, M., & Sadato, N. (2004). The neural substrates of conscious color perception demonstrated using fMRI. *NeuroImage, 21*, 1665–1673.

Mountcastle, V. (1957). Modality and topographic properties of single neurons in cats somatic sensory cortex. *Journal of Neurophysiology, 20*, 408–434.

Mountcastle, V. (2003). Introduction. *Cerebral Cortex, 13*, 2–4.

Murphy, E. H., & Berman, N. (1979). The rabbit and the cat: A comparison of some features of response properties of single cells in the primary visual cortex. *Journal of Comparative Neurology, 188*, 401–427.

Murray, S. O., & He, S. (2001). Contrast invariance in the human lateral occipital complex depends on attention. *Cerebral Cortex, 16*, 606–611.

Nair-Collins, M. (2013). Representation in biological systems: Teleofunction, etiology, and structural preservation. In L. Swan (Ed.), *Origins of mind* (pp. 161–185). New York: Academic.

Nealey, T. A., & Maunsell, J. H. R. (1994). Magnocellular and parvocellular contributions to the responses of neurons in macaque striate cortex. *Journal of Neuroscience, 14*, 2069–2079.

Nessler, B., Pfeiffer, M., Buesing, L., & Maass, W. (2013). Bayesian computation emerges in generic cortical microcircuits through spike-timing-dependent plasticity. *PLoS Computational Biology, 9*, e1003037.

Nestor, A., Plaut, D. C., & Behrmann, M. (2011). Unraveling the distributed neural code of facial identity through spatiotemporal pattern analysis. *Proceedings of the Natural Academy of Science USA, 108*, 9998–10003.

Newell, A., & Simon, H. A. (1976). Computer science as empirical enquiry: Symbols and search. *Communications of the Association for Computing Machinery, 19*, 113–126.

Niell, C., & Stryker, M. P. (2010). Modulation of visual responses by behavioral state in mouse visual cortex. *Neuron, 65*, 472–479.

Nieuwenhuys, R., Voogd, J., & van Huijzen, C. (2008). *The human central nervous system*. Berlin: Springer.

O'Brien, G., & Opie, J. (2004). Notes toward a structuralist theory of mental representation. In H. Clapin, P. Staines, & P. Slezak (Eds.), *Representation in mind – New approaches to mental representation*. Amsterdam: Elsevier.

O'Keefe, J., & Nadel, L. (1978). *The hippocampus as a cognitive map*. Oxford: Oxford University Press.

O'Keefe, J., & Recce, M. (1993). Phase relationship between hippocampal place units and the EEG theta rhytm. *Hippocampus, 3*, 317–330.

Orban, G. A., Zhu, Q., & Vanduffel, W. (2014). The transition in the ventral stream from feature to real-world entity representations. *Frontiers in Psychology, 5*, 695.

Osterhout, L., Kim, A., & Kuperberg, G. R. (2007). The neurobiology of sentence comprehension. In M. Spivey, K. McRae, & M. Joanisse (Eds.), *The Cambridge handbook of psycholinguistics* (pp. 365–389). Cambridge: Cambridge University Press.

O'Toole, A. J., Jiang, F., Abdi, H., & Haxby, J. V. (2005). Partially distributed representations of objects and faces in ventral temporal cortex. *Journal of Cognitive Neuroscience, 17*, 580–590.

Palmer, S. (1978). Fundamental aspects of cognitive representation. In E. Rosch & B. Lloyd (Eds.), *Cognition and categorization*. Mahwah: Lawrence Erlbaum Associates.

Papeo, L., Negri, G. A. L., Zadini, A., & Rumiati, R. I. (2010). Action performance and action-word understanding: Evidence of double dissociations in left-damaged patients. *Cognitive Neuropsychology, 27*, 428–461.

Papineau, D. (1987). *Reality and representation*. Oxford: Basil Blackwell.

Papineau, D. (1993). *Philsophical naturalism*. Oxford: Basil Blackwell.

Paradiso, M., MacEvoy, S. P., Huang, X., & Blau, S. (2005). The importance of modulatory input for V1 activity and perception. *Progress in Brain Research, 149*, 257–267.

Pavlov, I. (1927). *Conditioned reflexes*. Oxford: Oxford University Press.

Peelen, M. V., & Downing, P. E. (2007). The neural basis of visual body perception. *Nature Reviews Neuroscience, 8*, 636–648.

Piccinini, G. (2006). Computational explanation in neuroscience. *Synthese, 153*, 343–353.

Piccinini, G. (2007). Computational modeling vs. computational explanation: Is everything a Turing Machine, and does it matter to the philosophy of mind? *Australasian Journal of Philosoph, 85*, 93–115.

Piccinini, G., & Bahar, S. (2013). Neural computation and the computational theory of cognition. *Cognitive Science, 34*, 453–488.

Piccinini, G., & Scarantino, A. (2010). Computation vs. information processing: Why their difference matters to cognitive science. *Studies in History and Philosophy of Science, 41*, 237–246.

Plate, T. (2003). *Holographic reduced representations*. Stanford: CSLI Publication.

Plebe, A. (2008). The ventral visual path: Moving beyond V1 with computational models. In T. A. Portocello & R. B. Velloti (Eds.), *Visual cortex: New research* (pp. 97–160). New York: Nova Science Publishers.

Pour-El, M. B., & Richards, I. (1981). Wave equation with computable initial data such that its unique solution is not computable. *Advances in Mathematics, 39*, 215–239.
Press, W. A., Brewer, A. A., Dougherty, R. F., Wade, A. R., & Wandell, B. A. (2001). Visual areas and spatial summation in human visual cortex. *Vision Research, 41*, 1321–1332.
Prinz, J. (2002). *Furnishing the mind – Concepts and their perceptual basis*. Cambridge: MIT.
Prinz, J. (2006b). Is the mind really modular? In R. Morris & L. Tarassenko (Eds.), *Cognitive systems: Information processing meets brain science* (pp. 22–36). Amsterdam: Elsevier.
Pulvermüller, F. (2010). Brain embodiment of syntax and grammar: Discrete combinatorial mechanisms spelt out in neuronal circuits. *Brain and Language, 112*, 167–179.
Pulvermüller, F., & Fadiga, L. (2010). Active perception: Sensorimotor circuits as a cortical basis for language. *Nature Reviews Neuroscience, 11*, 351–360.
Putnam, H. (1988). *Representation and reality*. Cambridge: MIT.
Pylyshyn, Z. (1981). Computation and cognition: Issues in the foundations of cognitive science. *Behavioral and Brain Science, 3*, 111–150.
Rajimehr, R., Devaney, K. J., Bilenko, N. Y., Young, J. C., & Tootell, R. B. (2011). The "parahippocampal place area" responds preferentially to high spatial frequencies in humans and monkeys. *PLoS Computational Biology, 9*, e1000608.
Rakic, P. (2008). Confusing cortical columns. *Proceedings of the Natural Academy of Science USA, 34*, 12,099–12,100.
Ramsey, W. M. (2007). *Representation reconsidered*. Cambridge: Cambridge University Press.
Rauschecker, J. P., & Scott, S. K. (2009). Maps and streams in the auditory cortex: Nonhuman primates illuminate human speech processing. *Nature Neuroscience, 12*, 718–724.
Ray, E., & Heyes, C. (2011). Imitation in infancy: The wealth of the stimulus. *Developmental Science, 14*, 92–105.
Rescorla, R. A., & Wagner, A. R. (1972). A theory of pavlovian conditioning: Variations in the effectiveness of reinforcement and nonreinforcement. In A. H. Black & W. F. Prokasy (Eds.), *Classical conditioning II: Current theory and research* (pp. 64–99). New York: Appleton Century Crofts.
Rice, H. (1954). Classes of recursively enumerable sets and their decision problems. *Transaction of American Mathematical Society, 74*, 358–366.
Rizzolatti, G., & Arbib, M. A. (1998). Language within our grasp. *Trends in Neuroscience, 21*, 188–194.
Rizzolatti, G., & Matelli, M. (2003). Two different streams form the dorsal visual system: Anatomy and functions. *Experimental Brain Research, 153*, 146–157.
Rizzolatti, G., & Sinigaglia, C. (2010). The functional role of the parieto-frontal mirror circuit: Interpretations and misinterpretations. *Nature Reviews Neuroscience, 11*, 264–274.
Rizzolatti, G., Camarda, R., Fogassi, L., Gentilucci, M., Luppino, G., & Matelli, M. (1988). Functional organization of inferior area 6 in the macaque monkey. *Experimental Brain Research, 71*, 491–507.
Rizzolatti, G., Fogassi, L., & Gallese, V. (2001). Neurophysiological mechanisms underlying the understanding and imitation of action. *Nature Reviews Neuroscience, 2*, 661–670.
Robbins, R. A., & McKone, E. (2007). No face-like processing for objects-of-expertise in three behavioural tasks. *Cognition, 103*, 34–79.
Rockel, A., Hiorns, R., & Powell, T. (1980). The basic uniformity in structure of the neocortex. *Brain, 103*, 221–244.
Rodieck, R. W. (1965). Quantitative analysis of cat retinal ganglion cell response to visual stimuli. *Vision Research, 5*, 583–601.
Rogalsky, C., Love, T., Driscoll, D., Andersond, S. W., & Hickok, G. (2011). Are mirror neurons the basis of speech perception? Evidence from five cases with damage to the purported human mirror system. *Neurocase, 17*, 178–187.
Romanski, L. M., Tian, B., Fritz, J., Mishkin, M., Goldman-Rakici, P. S., & Rauschecker, J. (1999). Dual streams of auditory afferents target multiple domains in the primate prefrontal cortex. *Nature Neuroscience, 2*, 1131–1136.

Rose, D. (1979). Mechanisms underlying the receptive field properties of neurons in cat visual cortex. *Vision Research, 19*, 533–544.

Rubenstein, J. L. R., Rakic, P. (Eds.). (2013a). *Comprehensive developmental neuroscience: Neural circuit development and function in the healthy and diseased brain*. New York: Academic.

Rubenstein, J. L. R., Rakic, P. (Eds.). (2013b). *Comprehensive developmental neuroscience: Patterning and cell type specification in the developing CNS and PNS*. New York: Academic.

Rupert, R. D. (1999). The best test theory of extension: First principle(s). *Minds and Language, 14*, 321–355.

Russell, B. (1927). *The analysis of matter*. London: Harcourt.

Ryder, D. (2004). SINBAD neurosemantics: A theory of mental representation. *Minds and Machines, 19*, 211–240.

Ryder, D. (2009a). Problems of representation I: Nature and role. In J. Symons & P. Calvo (Eds.), *The Routledge companion to philosophy of psychology* (pp. 233–250). London: Routledge.

Ryder, D. (2009b). Problems of representation II: Naturalizing content. In J. Symons & P. Calvo (Eds.), *The Routledge companion to philosophy of psychology* (pp. 251–279). London: Routledge.

Saur, D., Kreher, B. W., Schnell, S., Kümmerer, D., Kellmeyer, P., Vry, M. S., Umarova, R., Musso, M., Glauche, V., Abel, S., Huber, W., Rijntjes, M., Hennig, J., & Weiller, C. (2008). Ventral and dorsal pathways for language. *Proceedings of the Natural Academy of Science USA, 105*, 18035–18040.

Schneider, G. E. (1967). Visual receptors and retinal interaction. *Science, 164*, 270–278.

Schouenborg, J., Garwicz, M., & Danielsen, N. (Eds.). (2011). *Brain machine interfaces – Implications for science, clinical practice and society*. Amsterdam: Elsevier.

Searle, J. R. (1990). Is the brain a digital computer? *Proceedings and Addresses of the American Philosophical Association, 64*, 21–37.

Shagrir, O. (2012). Structural representations and the brain. *British Journal for the Philosophy of Science, 63*, 519–545.

Shanks, D. R. (1995). *The psychology of associative learning*. Cambridge: Cambridge University Press.

Shepard, R. N., & Chipman, S. (1970). Second-order isomorphism of internal representations: Shapes of states. *Cognitive Psychology, 1*, 1–17.

Shepherd, G. M. (1988). A basic circuit for cortical organization. In M. S. Gazzaniga (Ed.), *Perspectives on memory research* (pp. 93–134). Cambridge: MIT.

Simons, J. S., Koutstaal, W., Prince, S., Wagner, A. D., & Schacter, D. L. (2003). Neural mechanisms of visual object priming: Evidence for perceptual and semantic distinctions in fusiform cortex. *NeuroImage, 19*, 613–626.

Singer, W. (1995). Synchronization of neuronal responses as a putative binding mechanism. In M. Arbib (Ed.), *The handbook of brain theoriey and neural networks* (pp. 960–964). Cambridge: MIT.

Sirosh, J., Miikkulainen, R., & Choe, Y. (Eds.). (1996). *Lateral interactions in the cortex: Structure and function*. Austin: The UTCS Neural Networks Research Group.

Slezak, P. (2002). The tripartite model of representation. *Philosophical Psychology, 15*, 239–270.

Smith, B. C. (2002). The foundations of computing. In M. Scheutz (Ed.), *Computationalism – New directions* (pp. 23–58). Cambridge: MIT.

Sprevak, M. (2011). William M. Ramsey, representation reconsidered. *British Journal for the Philosophy of Science, 62*, 669–675.

Stettler, D. D., Das, A., Bennett, J., & Gilbert, C. D. (2002). Lateral connectivity and contextual interactions in macaque primary visual cortex. *Neuron, 36*, 739–750.

Stowe, L. A., Haverkort, M., & Zwarts, F. (2004). Rethinking the neurological basis of language. *Lingua, 115*, 997–1042.

Swindale, N. V. (2001). Keeping the wires short: A singularly difficult problem. *Neuron, 29*, 316–317.

Swoyer, C. (1991). Structural representation and surrogative reasoning. *Synthese, 87*, 449–508.

Tanaka, K. (1996). Inferotemporal cortex and object vision. *Annual Review of Neuroscience, 19*, 109–139.

Tanaka, K., Saito, H., Fukada, Y., & Moriya, M. (1991). Coding visual images of objects in the inferotemporal cortex of the macaque monkey. *Journal of Neurophysiology, 66*, 170–189.

Tarr, M. J., & Bülthoff, H. H. (1998). Image-based object recognition in man, monkey, and machine. In M. J. Tarr & H. H. Bülthoff (Eds.), *Object recognition in man, monkey, and machine*. Cambridge: MIT.

Tarr, M., & Gauthier, I. (2000). FFA: A flexible fusiform area for subordinate-level visual processing automatized by expertise. *Nature Neuroscience, 3*, 764–769.

Tenenbaum, J. B., Kemp, C., Griffiths, T. L., & Goodman, N. D. (2011). How to grow a mind: Statistics, structure, and abstraction. *Science, 331*, 1279–1285.

Tettamanti, M., & Moro, A. (2012). Can syntax appear in a mirror (system)? *Cognition, 48*, 923–935.

Thivierge, J. P., & Marcus, G. F. (2007). The topographic brain: From neural connectivity to cognition. *Trends in Neuroscience, 30*, 251–259.

Thorndike, E. (1892). Animal intelligence: An experimental study of the associative processes in animals. *Psychological Monographs, 2*, 192–205.

Tiao, Y., & Blakemore, C. (1976). Functional organization in the visual cortex of the golden hamster. *Journal of Comparative Neurology, 168*, 459–481.

Toni, I., de Lange, F. P., Noordzij, M. L., & Hagoort, P. (2008). Language beyond action. *Journal of Physiology – Paris, 102*, 71–79.

Tootell, R. B., Silverman, M. S., Hamilton, S. L., Switkes, E., & De Valois, R. (1988a). Functional anatomy of the macaque striate cortex. V. spatial frequency. *Journal of Neuroscience, 8*, 1610–1624.

Tootell, R. B., Switkes, E., Silverman, M. S., & Hamilton, S. L. (1988b). Functional anatomy of the macaque striate cortex. I. Ocular dominance, binocular interactions, and baseline conditions. *Journal of Neuroscience, 8*, 1531–1568.

Tootell, R. B., Switkes, E., Silverman, M. S., & Hamilton, S. L. (1988c). Functional anatomy of the macaque striate cortex. II. Retinotopic organization. *Journal of Neuroscience, 8*, 1531–1568.

Tootell, R. B., Switkes, E., Silverman, M. S., & Hamilton, S. L. (1988d). Functional anatomy of the macaque striate cortex. III. Color. *Journal of Neuroscience, 8*, 1531–1568.

Trevarthen, C., & Sperry, R. W. (1968). Two mechanisms of vision in primates. *Psychological Research, 31*, 299–337.

Tsien, J. Z. (2007). The organizing principles of real-time memory encoding: Neural clique assemblies and universal neural codes. In B. Bontempi, A. Silva, & Y. Christen (Eds.), *Memories: Molecules and circuits* (pp. 100–182). Berlin: Springer.

Turing, A. (1936) On computable numbers, with an application to the Entscheidungsproblem. *Proceedings of the London Mathematical Society, 42*, 230–265.

Umarova, R. M., Saur, D., Schnell, S., Kaller, C. P., Vry, M. S., Glauche, V., Rijntjes, M., Hennig, J., Kiselev, V., & Weiller, C. (2010). Structural connectivity for visuospatial attention: Significance of ventral pathways. *Cerebral Cortex, 20*, 121–129.

Umiltà, M., Kohler, E., Gallese, V., Fogassi, L., Fadiga, L., Keysers, C., & Rizzolatti, G. (2001). I know what you are doing – A neurophysiological study. *Neuron, 31*, 155–165.

Ungerleider, L., & Mishkin, M. (1982). Two cortical visual systems. In D. J. Ingle, M. A. Goodale, & R. J. W. Mansfield (Eds.), *Analysis of visual behavior* (pp. 549–586). Cambridge: MIT.

Usher, M. (2004). Comment on Ryder's SINBAD neurosemantics: Is teleofunction isomorphism the way to understand representations? *Minds and Language, 19*, 241–248.

Vanduffel, W., Tootell, R. B., Schoups, A. A., & Orban, G. A. (2002). The organization of orientation selectivity throughout the macaque visual cortex. *Cerebral Cortex, 12*, 647–662.

Van Essen, D. C. (2005). A population-average, landmark- and surface-based (PALS) atlas of human cerebral cortex. *NeuroImage, 28*, 635–662.

Van Essen, D. C., & DeYoe, E. A. (1994). Concurrent processing in the primate visual cortex. In M. S. Gazzaniga (Ed.), *The cognitive neurosciences*. Cambridge: MIT.

Van Essen, D. C., Newsome, W., & Maunsell, J. (1984). The visual field representation in striate cortex of the macaque monkey: Asymmetries, anisotropies, and individual variability. *Vision Research, 24*, 429–448.

Van Essen, D. C., Lewis, J. W., Drury, H. A., Hadjikhani, N., Tootell, R. B., Bakircioglu, M., & Miller, M. I. (2001). Mapping visual cortex in monkeys and humans using surface-based atlases. *Vision Research, 41*, 1359–1378.

van Fraassen, B. C. (2008). *Scientific representation*. Oxford: Oxford University Press.

Van Hooser, S. D., Heimel, J. A. F., Chung, S., Nelson, S. B., & Toth, L. J. (2005). Orientation selectivity without orientation maps in visual cortex of a highly visual mammal. *Journal of Neuroscience, 25*, 19–28.

Verkindt, C., Bertrand, O., Echallier, F., & Pernier, J. (1995). Tonotopic organization of the human auditory cortex: N100 topography and multiple dipole model analysis. *Electroencephalography and Clinical Neurophisiology, 96*, 143–156.

von der Malsburg, C. (1973). Self-organization of orientation sensitive cells in the striate cortex. *Kybernetic, 14*, 85–100.

von der Malsburg, C. (1995a). Binding in models of perception and brain function. *Current Opinion in Neurobiology, 5*, 520–526.

von Economo, C., & Koskinas, G. N. (1925). *Die Cytoarchitektonik der Hirnrinde des erwachsenen Menschen*. Berlin: Springer.

von Neumann, J. (1961). The general and logical theory of automata. In A. Taub (Ed.), *Collected works* (Vol. V, pp. 288–328). New York: Pergamon Press.

Vuilleumier, P., Henson, R. N., Driver, J., & Dolan, R. J. (2002). Multiple levels of visual object constancy revealed by event-related fMRI of repetition priming. *Nature Neuroscience, 5*, 491–499.

Wandell, B. A. (1999). Computational neuroimaging of human visual cortex. *Annual Review of Neuroscience, 10*, 145–173.

Wandell, B. A., Brewer, A. A., & Dougher, R. F. (2005). Visual field map clusters in human cortex. *Philosophical Transactions of the Royal Society of London, 360*, 693–707.

Weigelt, S., Kourtzi, Z., Kohler, A., Singer, W., & Muckli, L. (2007). The cortical representation of objects rotating in depth. *Journal of Neuroscience, 27*, 3864–3874.

Wiesel, T., & Hubel, D. (1965). Binocular interaction in striate cortex of kittens reared with artificial squint. *Journal of Neurophysiology, 28*, 1041–1059.

Williams, D., Sagness, K., & McPhee, J. (1994). Configural and elemental strategies in predictive learning. *Journal of Experimental Psychology, Learning, Memory, and Cognition, 20*, 694–709.

Winer, J. A., Miller, L. M., Lee, C. C., & Schreiner, C. E. (2005). Auditory thalamocortical transformation: Structure and function. *Neuron, 28*, 255–263.

Wittgenstein, L. (1953). *Philosophische Untersuchung*. Oxford: Basil Blackwell.

Wolfram, S. (2002). *A new kind of science*. Champaign: Wolfram Media.

Wunderlich, K., Symmonds, M., Bossaerts, P., & Dolan, R. J. (2011). Hedging your bets by learning reward correlations in the human brain. *Neuron, 71*, 1141–1152.

Xu, Y. (2005). Revisiting the role of the fusiform face area in visual expertise. *Cerebral Cortex, 15*, 1234–1242.

Zeki, S. (1971). Functional organization of a visual area in the posterior bank of the superior temporal sulcus of the rhesus monkey. *Journal of Physiology, 236*, 549–573.

Zeki, S. (1983a). Colour coding in the cerebral cortex: The reaction of cells in monkey visual cortex to wavelenghts and colours. *Neuroscience, 9*, 741–765.

Zeki, S. (1983b). Colour coding in the cerebral cortex: The responses of wavelength-selective and colour-coded cells in monkey visual cortex to changes in wavelength composition. *Neuroscience, 9*, 767–781.

Zeki, S. (2003). Improbable areas in the visual brain. *Trends in Neuroscience, 26*, 23–26.

# Chapter 4
# Modeling Neural Representations

**Abstract** As the final part of the semantics of neurons, and as a prelude to the second part, the neurosemantics of language, this chapter seeks mathematical formulations for the mechanisms that enable the construction of representations in the brain. It is not a general review of the rich variety of mathematical solutions proposed so far for simulating neural circuits, currently available on the market. It is the introduction to the mathematical framework adopted in all the neurosemantic models that will be described in the second part.

One of the main challenges any endeavor of mathematical formalization of neural activities must face, is their impressive abundance.

The number of neurons involved in almost all cognitive functions is so large that it is impossible to give an overall sense of their activity by means of individual descriptions. Mathematics only offers the great advantage of synthesis, the possibility of capturing in a concise formulation the principles ruling the behavior of millions of interacting elements. Mathematical formulations can be implemented in a software, and the simulated results can be analyzed in detail.

In the cortex, neurons are characterized not only by their large number, but also by properties such as local cooperative and competitive interactions, which fit well within an established mathematical framework, that of self-organization. The adopted neural architecture derives from this general framework. In the interpretation of the activities of many neurons in the same cortical area, resulting from a simulation, a well established neurocomputational concept will be used, that of population coding, discussed in the last section of this chapter.

## 4.1 Self Organizazion in the Cortex

Self-organization is a term that has met with much success, and as a result has pervaded a range of disciplines that are very different from each other, from biology, to economy, to sociology. This makes it necessary, after a general introduction of the concept, to restrict its meaning to a narrow interpretation of it as a computational principle useful in explaining processes that take place at a certain scale of neural circuitry.

 
A. Plebe, V.M. De La Cruz, *Neurosemantics*,
Studies in Brain and Mind 10, DOI 10.1007/978-3-319-28552-8_4

As far as we know, the term self-organization was first mentioned by Ashby (1947), who used it as a basic concept in the newly born field of cybernetics. Its purpose was to reconcile two apparently clashing facts found in biological systems or artificial machines, that of being strictly determined in their actions by physical-chemical processes, and yet able to undergo self-induced internal reorganizations resulting in changes of behavior. In one of his later papers (Ashby 1962) related the idea of "organization" to the complexity of the system, such as the dependence of the system's behavior upon a usually high number of interacting variables. It contrasts with a system in which variables can be separated in mathematical forms. For a system to develop a certain organization, under this definition, is rather trivial, and does not necessarily qualify as a case of "self-organization", which, for Ashby, is the property of changing from a "bad" to a "good" organization. However, there is no a priori criteria for evaluating the developed organization as being good in any absolute sense, not only does it depend on each specific system, it is also a property that is observable only from the outside, that is, only when the organization has been reached. In the case of a brain, for instance, an organization can be deemed "good" if it acts so as to provide some kind of advantage to the organism's survival.

One difficulty in isolating self-organization phenomena, for Ashby, consisted in the fact that ordinarily, the systems being dealt with were either too simple, such as a pendulum with very few variables and a single equilibrium, or too complex, such as a living organism. Suitable examples, however, became rapidly available in physics and chemistry. A striking case was observed as early as the beginning of the last century by the French physicist Henri Claude Bénard (1900), that of convective cells that organize in a fluid. When heating water in a pot, buoyancy produces the upwelling of lesser dense molecules, which for mass conservation should be compensated by the downward motion of colder molecules. While initially the billions of water molecules exhibit a random motion, gradually, a small number of regular patterns of cells organizes. The cells are formed by an inner cylinder with a laminar upward flow, bounded by downward flow at the periphery. We usually miss this amazing phenomena because water is transparent, it can be experimentally visualized using solid markers, as in the middle image of Fig. 4.1.

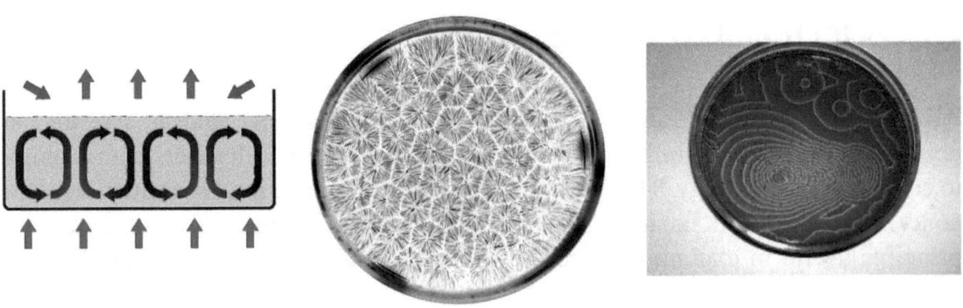

**Fig. 4.1** Examples of self-organization in physics and chemistry. On the left a scheme of Bénard's convective cells, with the heated fluid directed upwards, and the coldest part directed downwards. In the middle a real image, with heated silicon oil, and the cells made visible thanks to graphite marker. On the right the B-Z reaction

Something very similar happens in chemical reactions. The B-Z reaction, for example, is well known, named by its discoverers Belousov (1959) and Zhabotinsky (1964). It is a mix of potassium bromate, cerium(IV) sulfate, malonic acid and citric acid in dilute sulfuric acid. The malonic acid reduces cerium(IV) ions into cerium(III), which in turn tends to be oxidized back to cerium(IV) ions by the potassium bromate. Here again, the alternate disposition of the two types of cerium ions is initially random, but gradually tends to organize into macroscopic patterns, like those shown at the right in Fig. 4.1. In the 1960s Haken (1978) initiated a research program, called "synergetics", to construct an unifying mathematical framework within which quantitative descriptions of self-organizing physical, chemical, and even biological systems can be made. One of the pivotal equations in this framework is the Fokker-Planck equation, which comes in several formulations, for example:

$$\frac{\partial}{\partial t}p(\mathbf{q}, t) = -\nabla_{\mathbf{q}}(p\mathbf{k}) \tag{4.1}$$

describes the evolution in time of the probability density $p$ for a point $\mathbf{q}$ in a space, for example the three geometrical coordinate of the position of a molecule, and $k$ is a velocity field in the space of $\mathbf{q}$. This equation statistically links the microscopic level of the motion, or other general characteristics, of the elementary components of a system, with its mesoscopic level. The field of synergetics is clearly closely related to the mathematics of dynamic systems in nonequilibrium phase transitions, and chaos theory.

### *4.1.1 Relationship with the Concept of Emergence*

Even more well known than self-organization is the term "emergence", which is often used interchangeably. Emergent properties spring up somehow from the interactions between the local parts of the system. Emergentism in fact, has a longer tradition, and was established as a specific stream of thought in philosophy of science during the late-nineteenth century by a group of British scholars, which included John Stuart Mill, was mostly represented by Charlie Dunbar Broad (1925). At that time the idea of emergentism was mainly a compromise in the heated debate between the mechanistic view, that living organisms were governed by the same physical-chemical principles of inorganic matter, and the vitalist position, that instead posited a kind of "vital substance" unique to living organisms. Emergentists confute the latter position, still retaining that the phenomena of life cannot be reduced to the effects of the component substances of the organism. According to Broad, the reason is that matter aggregates at different levels, and each is characterized by certain irreducible properties that "emerge" from lower-level properties.

Over the years, the impact of emergence declined in philosophy of science, with interest waning primarily due to the remarkable success of mechanistic explanations for many of the properties of life that have been advanced in the last century. The concept of emergence, however, has found a renewed vigor these past few decades, in large part thanks to the theoretical and mathematical development of self-organization, leading to a sort of "re-emergence of emergence" (Clayton and Davies 2006).

From a mathematical point of view, emergentism is not confined to the self-organization framework, as developed in synergetics, it embraces different directions of research, such as the study of dissipative structures at far-from-equilibrium conditions, started by Ilya Prigogine (1961), or the theory of attractors in chaos theory, first proposed by Ruelle and Takens (1971). Within the contemporary scene, two broad positions can be distinguished, between the proposers of an epistemological emergence, and the supporters of an ontological conception of emergence, more resilient than the British version of emergentism.

Clearly, the shift from ontological emergence to plain Cartesian dualism is easy, for example Hasker (1999) plainly tries to defend a kind of dualism, he dubs "emergent dualism". Ontological emergentism, therefore, is in danger of falling prey to the well known difficulties of reconciling dualism with the unity of the physical world, as known in contemporary science. In addition, it suffers from specific and serious metaphysical weaknesses, probably the most threatening objection is Kim (2006)'s downward causation argument. According to emergentism, a property $P$ observable at mesoscopic level, is always the result of a set $\mathscr{B} = \{B_1, B_2, \cdots\}$ of basic properties $B_i$, that take place at a microscopic level. Now suppose property $P$ causes a different "emerged" property $\tilde{P}$. By definition, the theory requires the existence of a different set $\tilde{\mathscr{B}} = \{\tilde{B}_1, \tilde{B}_2, \cdots\}$ with the low-level properties on which $\tilde{P}$ supervenes. The puzzle is that, in order to save the causation from $P$ to $\tilde{P}$, it is necessary to also postulate a "downward" causation from $P$ to the lower set $\tilde{\mathscr{B}}$, which is highly problematic. On the other hand, purifying emergentism from claims of causal effects at a mesoscopic level is not possible, because it is exactly what emergentists promise to explain. For example, in the case of the brain, consciousness is the primary exemplar of emergent phenomena, and there is an impressive demand of causal power to conscious thought, from manipulating the beliefs of others, to that of producing artifacts.

Our use of self-organization is more in line with epistemological emergence. For most exponents of epistemological emergence, however, it will encompass all kinds of phenomena that are striking from a macroscopic point of view. This stance is defended, for example, by Kauffman (1993, 2008), offering a wide range of mathematical formulations of self-organization, on which he speculates life to be based, and possibly the whole universe.

We instead interpret self-organization purely as a methodological tool, that may offer the benefit of mathematical synthesis, in certain cases were the number of interacting elements is very large. As reviewed by Willshaw (2006), in the nervous system many phenomena can be characterized as self-organizing, although there are very few cases of pure self-organization. In most of the observed phenomena the final state depends largely on external influences, either from other brain areas, or from sensory stimulation from the outside world. The most important forms of neural self-organization appears during the development of the nervous system. At the cortical level, there is substantial evidence that both its regionalization, and the positioning of cells in patterns inside regions, are the effect of several combined mechanisms, with a fundamental role in the afferent thalamocortical axons and in the short range interactions between cortical cells.

### 4.1.2 First Mathematical Descriptions

The first attempts to use the mathematical framework of self-organization to describe neural phenomena are attributed to von der Malsburg (1973) and Willshaw and von der Malsburg (1976), who addressed some of the most striking types of organization found in the cortex: the maps in the visual system (see Sect. 3.3.3). There are three key mechanisms in cortical circuits that match with the premises of self-organization:

1. small signal fluctuations might be amplified, this is a direct effect of the non-linear behavior of neurons;
2. there is cooperation between fluctuations, in that excitatory lateral connections tend to favor the firing of other connected neurons, and Hebbian law reinforces synapses of neurons that fire frequently in synchrony;
3. there is competition as well, in that inhibitory connections can lower the firing rate of groups of cells at the periphery of a dominant active group, and synaptic homeostasis compensates for the gain in contribution from more active cells, by lowering the synaptic efficiency of other afferent cells.

In the cortical model devised by von der Malsburg the activity $x_i$ of each neuron $i$ was computed by the following system of differential equations:

$$\frac{\partial}{\partial t}x_i(t) = -\alpha_i x_i(t) + \sum_{j\in\mathscr{C}_\rangle} w_{ij} f\left(x_j(t)\right) + \sum_{j\in\mathscr{A}_\rangle} w_{ij} a_j(t) \tag{4.2}$$

$$f\left(x_i(t)\right) = \begin{cases} x_i(t) - \theta_i & \text{if } x_i(t) > \theta_i \\ 0 & \text{otherwise} \end{cases} \tag{4.3}$$

where $\mathscr{C}_\rangle$ is the set of cortical neurons with lateral connections to the cell $i$, and $\mathscr{A}_\rangle$ is the set of all afferent axons, each carrying a signal $a(t)$. $w_{ij}$ are the synaptic efficiency between cell presynaptic $j$ and postsynaptic $i$, and are modified by an amount proportional to the presynaptic and postsynaptic signals, in the case of the coincidence of activity. Periodically all $w_{ij}$ leading to the same cortical cell $i$ are renormalized, realizing the competition, in that some synapses are increased at the expense of others.

The source of afferents, leading to the process of self-organization, can be the external scene seen by the eyes, but also spontaneous activity generated by the brain itself (Mastronarde 1983). The organization in the visual system explained by equations like those of (4.2), range from retinotopy, ocular dominance, to orientation sensitivity (von der Malsburg 1995b).

### 4.1.3 The SOM Algorithm

During the 1980s not many scholars were familiar with the early models of von der Malsburg. It was the period of the connectionist boom in neural computation,and the much simpler and efficient feedforward neural schemes (Rumelhart and McClelland 1986) became the preferred choice of researchers. The differential equations in (4.2), on the contrary, had no analytic solution, and required complex numerical integration, something that was computationally expensive at the time.

Soon a very different proposal arrived, offering self-organizing properties, like the von der Malsburg models, at a much cheaper price, with the same simplicity and efficiency of the other models in the connectionist arena. This proposal is known with the acronym SOM (*Self-Organizing features Map*), and also as Kohonen (1982, 1984, 1995) Maps, named after their proposer. The first term in the acronym reveals the ambition of this algorithm, that of implementing some sort of self-organization, the last term better specifies the form through which the organization is achieved: topologically, inside a *map*. As discussed in Sect. 3.3.3, the map arrangement is the most basic ordering principle of the cortex. In the SOM it becomes more general, in that there is no specific number of dimensions, although the two-dimensional case is the most common, and clearly the most suitable for simulating visual phenomena. Moreover, the SOM has been largely used for applications outside of brain modeling, and its main benefit is that of reducing high dimensional data in spaces where the relationship between data can be grasped, and the best final dimension, for humans, is two-dimensional. In evaluations on how close the SOM algorithm is to that of the real physiology of the cortex (von der Malsburg 1995b) the ensuing comments were not kind, defining Kohonen maps as "an algorithmic caricature of the [self-organization] mechanism".

The SOM has its mathematical roots in *Vector Quantization*, a method used in signal processing to approximate with a small number of vectors, called *codebook*, the probability density of a stochastic high dimensional vector (Linde et al. 1980).

Being $\mathbf{t} \in \mathbb{R}^N$ the data to analyze, a two dimensional Kohonen map is made up by $M$ neurons $\mathbf{x} \in \mathbb{R}^N$, with an associated two-dimensional coordinate $\mathbf{r} \in \{< [0, 1], [0, 1] >\} \subset \mathbb{R}^2$. When data are presented to the network, the same vector is available to all neurons in the map. The main strategy of the algorithm is the so-called *winner-take-all*, the singling out of just one neuron over all $M$, which best responds to that specific input. A scheme of this network is provided in Fig. 4.2.

In mathematical terms, for a give input $\mathbf{t}$ the winner neuron $\mathbf{x}_c$ is chosen by the following equation:

$$c = \arg \min_{i \in \{1,\ldots,M\}} \{\|\mathbf{t} - \mathbf{x}_i\|\} \tag{4.4}$$

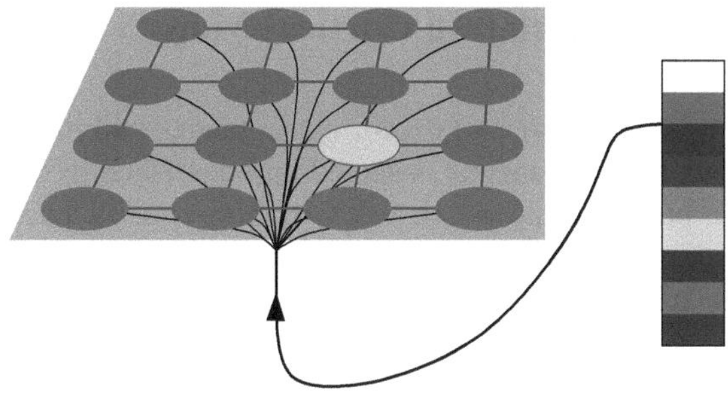

**Fig. 4.2** Scheme of a two dimensional Kohonen map. Every neuron receives the same vectorial input, as shown on the right. Only one neuron is the "winner", here marked with withe color

where the metrics for comparing two vectors is arbitrary. The same procedure is used during the learning phase of the network. At the beginning all neurons start with random vectors, and all samples $\mathbf{t} \in \mathscr{T}$ are presented in random order. After the selection of a winner neuron $c$, in response to a sample $\mathbf{t}$, using (4.4), the vectors associated with the neurons are modified, using this rule:

$$\Delta \mathbf{x}_i = \eta e^{-\frac{\|\mathbf{r}_c - \mathbf{r}_i\|^2}{2\sigma^2}} (\mathbf{t} - \mathbf{x}_i) \tag{4.5}$$

where $\eta$ is a the learning rate, and $\sigma$ is the width of the influence of the winner $c$ in adapting its neighbors. In (4.5) there are two components, this one

$$\eta\ (\mathbf{t} - \mathbf{x}_i) \tag{4.6}$$

acts in attracting $\mathbf{x}_i$ to the target $\mathbf{t}$, by an amount weighted by $\eta$. Equation (4.6) coincides with (4.5) in the case of the winner, $c = i$. For all the other neurons, an additional modulation is given by the second term of (4.5):

$$e^{-\frac{\|\mathbf{r}_c - \mathbf{r}_i\|^2}{2\sigma^2}} \tag{4.7}$$

limiting the vector update, as long as the neurons are far away from the winner, with a Gaussian shaped by $\sigma$. At the end of the learning phase, a relational ordering may come about in the map, so that the presentation of a $\mathbf{t}$ will trigger a winner neuron, in a spatial location of the map, relevant for some feature of the data.

Initially, this algorithm was only proven to successfully self-organize in several experiments, without being able to provide a mathematical demonstration. Later Cottrell and Fort (1987) gave a demonstration of the self-organization capability in the one-dimensional case. In the two-dimensional arrangement, Erwin et al. (1992a,b) established conditions that ensure the convergence to an ordered state. The SOM algorithm has been successfully applied in a range of simulations, from somatosensorial perception, motor control (Ritter et al. 1992), to abstract combinatorial optimization (Plebe and Anile 2001; Plebe 2001).

## 4.2 Simulating Cortical Maps

In the enterprise of exploring semantics within the framework of the set of essential neural mechanisms, selected and described in Chap. 3, the SOM algorithm will be a poor starting point. Even if some models based on SOM have been amazingly able to reproduce aspects of functional organizations found in neural maps (Yu et al. 2005), its algorithmic formulation is alien to biological processes. Units in a SOM map, even if sometimes called "neurons", cannot be related to any particular brain structure, neither neurons nor cortical columns. The success SOMs have in predicting self-organization phenomena in the brain is undoubtedly due – in part – to similarities between its mathematical formulation, and real brain mechanisms. The winner-take-all strategy broadly captures the effects of competitive neural inhibition, and the SOM adaptation rules somehow takes into account local neural interactions. However, it is a pure predictive model, that departs from the model-mechanism-mapping criteria discussed in Sect. 3.1.5. Therefore, the SOM is unable to account for the mechanisms listed in Chap. 3, selected as the basis for modeling neurosemantics.

Since the late 1990s a number of simulators have been developed, that include most of the details involved in the electrical behavior of neural cells, mostly following the Hodgkin-Huxley equations (see Sect. 2.1.2). The best known of these simulators are GENESIS (Bower and Beeman 1998) and NEURON (Hines and Carnevale 1997), with the latter having been adopted as the basis of the Blue Brain Project (Markram 2006), the attempt to reproduce large-scale brain circuits on massive parallel computers. Although modeling networks of realistic multi-compartment neurons is currently the best example of mechanistic neural simulations, and the unprecedented resources invested today, both in the States and in Europe, in large-scale brain models will certainly play a crucial role in future neuroscience, it is not necessarily the best solution for the neurosemantic enterprise. The premise that the more detailed the neural model is, the better the explanation it will offer, independently from the function to be investigated, is arguable (Eliasmith and Trujillo 2014).

For what concerns the specific goal here pursued, the details at the single cell level are not the object of inquiry, even if the studied behavior depends on all such details. The best strategy, we believe, is to adopt a level of modeling where the mechanisms at a lower level are included in an integrated way, and the interacting components of the model keep a plausible relationship with the relevant components of the neural system. The precise level of grain of analysis required to explain specific semantic phenomena remains, however, an open question.

Most of the models of semantics that will be described in the second part of this book have been developed using the Topographica system (Bednar 2009, 2014), chosen as one of the best compromises between the inclusion of all the essential mechanisms deemed responsible for neurosemantics, yet still simple enough to allow a reconstruction of a hierarchy of cortical areas relevant for vision and language processing. It has been specifically developed for modeling maps in the cortex, and its units may correspond to cortical columns. Its main features will be described in the sections that follow, and its mathematical formulations will be articulated.

### *4.2.1 Lateral Connections, Competitive Normalization*

Topographica is built on a previous model, named LISSOM (*Laterally Interconnected Synergetically Self-Organizing Map*) (Sirosh and Miikkulainen 1997), which brings together concepts of self-organization, map topology, and lateral connections, as the acronym reveals, and includes the mathematical framework of synergetics (Haken 1978). More precisely, the neural mechanisms included in LISSOM are the following:

1. the intercortical connections of inhibitory and excitatory types;
2. the afferent connections, of thalamic nature, or incoming from lower cortical areas;
3. the organization on two dimensions of neural coding;
4. the reinforcement of synaptic efficiency by Hebbian learning;
5. homeostatic compensation of neural excitability.

The first two points are consistent with the main properties of the cortex, as seen in Sect. 2.3.2, the third point hinges on one of the main organization principles of the cortex described in Sect. 3.3.3. The fourth point is the implementation of the coincidence detection principle proposed in Sect. 3.2, as the main mechanism for coding representations in neural circuits. The last point implements the mechanism of synaptic homeostasis, already included by von der Malsburg (1973) in his first mathematical attempt to use the mathematical framework of self-organization for neural circuits (see Sect. 4.1.2), which increasingly appears to be a prominent factor in refining synaptic connectivity (Turrigiano and Nelson 2004). Recently, it has been included as an instance of the more widespread normalization process defined as "canonical neural computation" (Carandini and Heeger 2012), for its presence in a diversity of neural systems, from the olfactory system of invertebrates, to the primary visual cortex, to higher visual and non-visual cortical areas. According to Chirimuuta (2014), it is an exemplar case of computational explanations peculiar to neuroscience, which escape the standard criteria of mechanistic explanation, while still providing compelling insights of a widespread neural process.

The LISSOM, being a simplified model of cortical areas, necessarily departs from its full-fledged behavior in several respects. Two main reasons of divergence are the following:

1. the electrical processes in the compartments of a single neuron are neglected, and there is no explicit generation of spikes in real time;
2. there is no differentiation of neural cell types inside the cortical columns.

For what concerns the first point, avoiding a model based on Hodgkin-Huxley equations, brings the considerable advantage of limiting the number of parameters describing every single neuron, which would have introduced useless degrees of variation in the overall system. The adopted solution is to use a scalar value, instead of the generation of action potential trains, in the conventional interval $[0 \cdots 1]$, representing the average frequency of the action potentials. There is a significant amount of experimental evidence on the high correlation between this synthetic value and the extant neural activation, in motor control (Milner-Brown et al. 1973), sensorial perception (Hubel and Wiesel 1962, 1968), and in vitro studies (de la Rocha et al. 2007). Nevertheless, this is a rather crude simplification, that unavoidably discards certain aspects of neural coding, such as the binding relations between distant areas (von der Malsburg 1995a; Singer 1995), discussed in Sect. 3.2.

The second point missing, concerns the differentiation of cell types in the cortex. As seen in Sect. 2.3, the cerebral cortex is populated by a variety of cells, distributed along its layers, this detail is missing in the LISSOM. This simplification, again, relieves the burden of having a large number of additional parameters, which are not of interest in the present investigation, but would have drastically increased the degrees of freedom of the system. A LISSOM unit, even if sometimes called "neuron", corresponds to a functional cortical aggregation, responding uniformly to an afferent signal, like the microcolumns seen in Sect. 3.3.1.

The basic equation in the LISSOM describes the activation level $x_i$ of the $i$-th unit in the map:

$$x_i^{(k)} = f\Bigg(\frac{\gamma_{\mathrm{A}}}{1 + \gamma_{\mathrm{N}}\mathbf{U} \cdot \mathbf{v}_{r_{\mathrm{A}},i}} \mathbf{a}_{r_{\mathrm{A}},i} \cdot \mathbf{v}_{r_{\mathrm{A}},i} + \gamma_{\mathrm{E}}\mathbf{e}_{r_{\mathrm{E}},i} \cdot \mathbf{x}_{r_{\mathrm{E}},i}^{(k-1)} - \gamma_{\mathrm{I}}\mathbf{i}_{r_{\mathrm{I}},i} \cdot \mathbf{x}_{r_{\mathrm{I}},i}^{(k-1)}\Bigg). \tag{4.8}$$

Vector $\mathbf{v}_{r_{\mathrm{A}},i}$ is composed by afferent to unit $i$ in a circular radius $r_{\mathrm{A}}$, the vectors $\mathbf{x}_{r_{\mathrm{E}},i}^{(k-1)}$ and $\mathbf{x}_{r_{\mathrm{I}},i}^{(k-1)}$ are the activation of all neurons in the map, where a lateral connection exists with neuron $i$ of an excitatory or inhibitory type, respectively. Their fields are circular areas of radius, respectively, $r_{\mathrm{E}}$, $r_{\mathrm{I}}$. Vectors $\mathbf{e}_i$ and $\mathbf{i}_i$ are composed by all connection strengths of the excitatory or inhibitory neurons projecting to $i$. The scalars $\gamma_{\mathrm{X}}$, $\gamma_{\mathrm{E}}$, and $\gamma_{\mathrm{I}}$, are constants modulating the overall contribution of afferents, excitatory, and inhibitory components.

The scalar $\gamma_{\mathrm{N}}$ controls the setting of a push-pull effect in the afferent weights, allowing inhibitory effects without negative weight values. Mathematically, it represents dividing the response from the excitatory weights by the response from a uniform disc of inhibitory weights over the receptive field of neuron $i$. In Eq. (4.8) and all the ones following the operation $\mathbf{x} \cdot \mathbf{y}$ are the product of vectors $\mathbf{x}$ and $\mathbf{y}$. Vector $\mathbf{U}$ is just a vector of 1's of the same dimension of $\mathbf{x}_i$. The function $f$ can be any monotonic nonlinear continuous growing function limited between 0 and 1, it is generally a piecewise linear approximation of the sigmoid function.

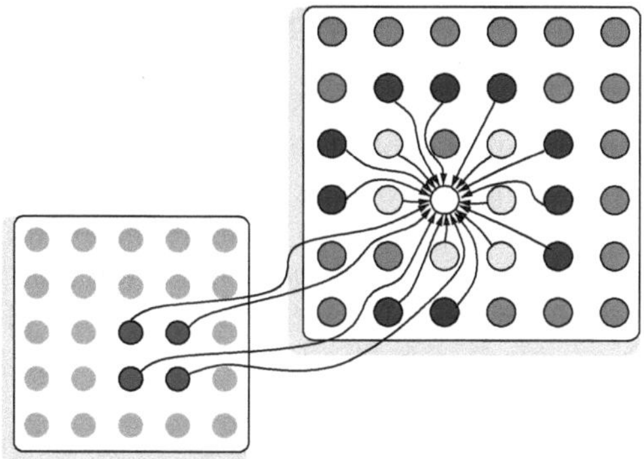

**Fig. 4.3** Scheme of connections of a unit in the LISSOM architecture. The unit in white color receives excitatory connections, in dark grey, inhibitory connections, in light grey, with additional afferences from thalamic maps, on the left

Equation (4.8) is recursive in time $k$, with initial condition $x_i^{(k=0)} = 0$, the final activation is defined as $x_i = x_i^{(k=K)}$, where $K$ satisfies the following condition:

$$\sum_i \left| x_i^{(k=K)} - x_i^{(k=K-1)} \right| < \epsilon \tag{4.9}$$

with $\epsilon$ a small defined value. A scheme of connections to a LISSOM unit is provided in Fig. 4.3. One of the main features that make the LISSOM scheme suitable for modeling the cortex is its inclusion of the computational contributions of intracortical lateral connections. A large number of physiological studies have confirmed anatomically a concentric pattern of excitatory and inhibitory lateral connectivity, extending densely roughly 500 μm around a center, and more patchy at longer distances (Gilbert et al. 1990; Grinvald et al. 1994; Sirosh et al. 1996; Stettler et al. 2002; Hou et al. 2003; Cerreira-Perpiñán and Goodhill 2004; Hunt et al. 2011).

All connections are plastic, and change in time according to the following rules:

$$\Delta \mathbf{a}_{r_{\mathrm{A}},i} = \frac{\mathbf{a}_{r_{\mathrm{A}},i} + \eta_{\mathrm{A}} x_i \mathbf{v}_{r_{\mathrm{A}},i}}{\|\mathbf{a}_{r_{\mathrm{A}},i} + \eta_{\mathrm{A}} x_i \mathbf{v}_{r_{\mathrm{A}},i}\|} - \mathbf{a}_{r_{\mathrm{A}},i}, \tag{4.10}$$

$$\Delta \mathbf{e}_{r_{\mathrm{E}},i} = \frac{\mathbf{e}_{r_{\mathrm{E}},i} + \eta_{\mathrm{E}} x_i \mathbf{x}_{r_{\mathrm{E}},i}}{\|\mathbf{e}_{r_{\mathrm{E}},i} + \eta_{\mathrm{E}} x_i \mathbf{x}_{r_{\mathrm{E}},i}\|} - \mathbf{e}_{r_{\mathrm{E}},i}, \tag{4.11}$$

$$\Delta \mathbf{i}_{r_{\mathrm{I}},i} = \frac{\mathbf{i}_{r_{\mathrm{I}},i} + \eta_{\mathrm{I}} x_i \mathbf{x}_{r_{\mathrm{I}},i}}{\|\mathbf{i}_{r_{\mathrm{I}},i} + \eta_{\mathrm{I}} x_i \mathbf{x}_{r_{\mathrm{I}},i}\|} - \mathbf{i}_{r_{\mathrm{I}},i}, \tag{4.12}$$

where $\eta_{\{\mathrm{A,E,I}\}}$ are the learning rates for the afferent, excitatory, and inhibitory weights, and $\|\cdot\|$ is the $\mathrm{L}^1$-norm. In all equations the numerator represents the Hebbian increase of synaptic efficiency due to coincidental activation of presynaptic and postsynaptic units. The denominator acts as a counterbalancing compensation, that tends to keep the average excitability of the neuron constant, in the long term, and implements the synaptic homeostatic mechanism, discussed above.

The Hebbian component in Eq. (4.12) is a standard mathematical formulation that abstracts the biological mechanisms of Hebbian plasticity described in Sect. 2.2.2, for a review of variants see Gerstner and Kistler (2002). Early mathematical implementations of Hebbian learning were known to face a number of computational problems, in particular the instability of synaptic strengths, that tend to approach either zero or some maximum strength (Miller and MacKay 1994), and the lack of synaptic competition (Dayan and Abbott 2001, p.284). Neither one is an issue for LISSOM. First, the Hebbian component is combined with homeostasis, the denominator term in (4.12). But mostly, neurons are placed in a circuit, where they do not act in isolation, but are strongly influenced by lateral inhibition and excitation. Even though each neuron is adapting its own connections, the activities of other neurons modulate the learning, within the self-organization process.

### *4.2.2 A Mathematical Framework for Hierarchical Cortical Maps*

The LISSOM architecture was originally conceived for simulating visual area V1, in direct connection with the thalamus (Bednar 2002). It evolved into Topographica, which has had several extensions, making it a valuable tool for composing complex hierarchies of cortical maps. One of the extensions to the basic equation, necessary for simulating multiple cortical areas, is in this equation:

$$\begin{aligned} x_i^{(k)} = f\Big( & \gamma_A g_A \left(\mathbf{a}_{r_A,i} \cdot \mathbf{v}_{r_A,i}\right) + \gamma_B g_B \left(\mathbf{b}_{r_B,i} \cdot \mathbf{u}_{r_B,i}\right) \\ & + \gamma_E \mathbf{e}_{r_E,i} \cdot \mathbf{x}_{r_E,i}^{(k-1)} - \gamma_I \mathbf{i}_{r_I,i} \cdot \mathbf{x}_{r_I,i}^{(k-1)} \Big). \end{aligned} \tag{4.13}$$

as the contribution given by vector $\mathbf{u}_{r_B,i}$, a backprojection from a higher area, it is the collection of the activities of neurons in that area, that project back to neuron $i$, within a receptive field or radius $r_{B,i}$. Vector $\mathbf{b}_{r_B,i}$ is made up by the synaptic efficiency of those projections. In addition, the overall normalization is no longer a single parameter, but can be a generic additional function $g(\cdot)$.

Moreover, an important variant of the monotonic non-linear saturation function $f$ has been introduced, as a threshold function, with an adaptive threshold $\theta$, dependent on the average activity of the unit itself. The threshold is updated as follows:

$$\theta^{(k)} = \theta^{(k)} + \lambda \left(\bar{x} - \mu\right) \tag{4.14}$$

where $\bar{x}$ is a smoothed exponential average in time of the activity, and $\lambda$ and $\mu$ fixed parameters. This feature simulates the biological adaptation that allows the development of stable topographic maps organized by preferred retinal location and orientation (Stevens et al. 2013).

## 4.3 Population Coding in the Cortex

In the effort of making progress in the interpretation of contents in higher-level maps, where a direct relation with peripheral inputs is lost, research these past few decades has drawn on the idea that the power of representing information in cortical circuits lies in the combination of the activities of many columnar units. This concept is usually named "distributed coding" (Hinton et al. 1986), but is also known as "population coding" , "vector coding" and "state space representation", in the formulation by Paul Churchland (1989). The idea has actually been around for quite some time, but mainly as intuitions without a strong relation to neurological data, as in Pribram (1971), who suggested that brain representations are distributed in force of a supposed analogy with holograms. In the current interpretation of population coding, a higher level map may code for a kind of object or fact, and it is the concurrent level of firing of a population of cells in that map that represents a specific instance of the kind. Since the 1990s, several studies have quantified how distributed the response in higher cortical areas to set of stimuli in a similar class is. In Sakai et al. (1994), monkeys were trained to remember synthetic pictures, at least 59 cells out of 91 recorded, responded to more than one picture. Other experiments done with natural faces (Rolls and Tovee 1995; Abbott et al. 1996) confirmed that not single cells, but populations of cells are necessary to discriminate single stimuli. Pasupathy and Connor (2002) studied the population coding by 109 cells in area V4 of macaque monkeys, of curvatures and angular positions from 49 simple patterns. The coding was demonstrated by reconstructing mathematically the 49 patterns from the population responses. A different stream of research inside distributed coding, attempts to establish computationally, the reasons and advantages nature has had for adopting this representational strategy in the cortex (Hinton et al. 1986; Olshausen and Field 1996; Brunel and Nadal 1998).

Recently, research on population coding has progressed in the statistical tools used in analyzing responses from distributed cortical areas, such as representational similarity analysis introduced by Kriegeskorte (2009), and intrinsic methods, where no predefined labeling of the coded categories is required (Lehky et al. 2013). Using representational similarity analysis of fMRI signals, Chikazoe et al. (2014) found a distributed coding of affective valence in the orbitofrontal cortex, supporting a continuous dimension of positive-to-negative valence. A comprehensive review of current findings of population coding in the brain is in Quian Quiroga and Panzeri (2013).

### 4.3.1 Code Sparseness

One property that has gained attention, in characterizing the way populations of neurons code entities, is their *sparseness*, the fraction of neurons engaged in representing a token with respect to the overall population. As reviewed by Olshausen and Field (2004), there is evidence that information is often represented by a surprisingly small number of simultaneously active neurons out of a large population, and there are several possible theoretical reasons for that. There are studies showing that when coding is sparse enough, Hebbian learning becomes more efficient (Kanerva 1993), an additional rather obvious reason is energy saving. Lennie (2003) estimated that, for energetic reasons, no more than one over 50 cortical neurons can be active at the same time.

Measuring the actual sparseness of a given neural population turned out not to be so straightforward, several mathematical formulations have been proposed, reviewed in Olshausen and Field (2004). There are mainly two mathematical accounts of sparseness: *population sparseness* in reference to the activation of a fraction of neurons in a population during fixed small time windows, and *lifetime sparseness*, the measure of how sporadic the activity of a single neuron over time is. The main experimental difficulty is when sparseness is high: when neurons fire rarely they are likely to be missed by the investigator. However, there is convergent evidence that lower sensorial areas have more dense coding, and higher sparseness is found in higher amodal areas. By using statistical analysis over a pool of recordings in the inferior temporal and prefrontal cortex of monkeys, Meyers et al. (2008) assessed the sparseness of the coding for synthetic images of cats and dogs. They found that the 64 best coding cells provide a classification accuracy almost indistinguishable from that of the entire population of 256 cells, and even the best 16 were able to provide a reasonable accuracy.

Note that the extreme upper limit to the sparseness is when in a population of neurons just one neuron at the time is activated by an individual object or event. This is the idea famously lampooned as the "grandmother cell", after a story invented by Jerry Lettvin in 1969, for a lecture course at MIT (Gross 2002). Akakhi Akakievitch, an imaginary great neurosurgeon, was able to identify all cells in the brain carrying information about one's own mother, and by ablating all such cells freed Portnoy (the character of Philip Roth's novel *Portnoy's Complaint*) from his obsession with his mother. After this success, Akakievitch moved on to search for cells representing one's grandmother. The term become widespread in neuroscience, where it often plays the role of straw man for discussions about neural coding. This idea recently found renewed vigor, after remarkable studies revealed specialized tuning of very few cells. In particular, in a series of studies in human epilepsy patients with presurgical implantation of microelectrodes, Quian Quiroga et al. (2007) demonstrated surprisingly precise tuning of entorhinal neurons to a variety of familiar images, such as Saddam Hussein. Using results such as these, Bowers (2009) attempted to resurrect the grandmother cell idea, arguing that it is not so bizarre after all. However, the same authors of the studies used by Bowers criticized his conclusions as unwarranted (Quian Quiroga et al. 2008; Quian Quiroga and Kreiman 2010). They remarked that, if there is one and only one cell responding to a person or concept, the chances for the investigator to find such cell, out of a few hundred million in the medial temporal lobe, would be infinitesimal.

### 4.3.2 Assessing the Population Coding in Topographica

In this section a general method for analytically assessing the organization of population coding in Topographica maps will be described.
For this purpose a number of ancillary functions will be introduced.

$$x_i(s) : S \in \mathscr{S} \to \mathbb{R}^+; \quad s \in S \in \mathscr{S}. \tag{4.15}$$

This equation computes the activation $x$ of a generic unit $i$ in a Topographica map, in response to the presentation of the stimulus $s$ from the external world to the system. Note that compared with the basic equation of Topographica (4.13) the activity $x$ in Eq. (4.15) has no time subscript: it is the stable activation, after the recurrent equation (4.13) has settled. Also the stimulus $s$ is treated as static, how to deal with events in time will be shown in the second part of this book, for the models where it is required. Note that the stimulus need not be a direct afferent to the map, it should of course, influence the excitation of the given map, in order to be represented in the coding. Let us assume the stimulus $s$ to be an instance of a possible sensorial experience given by an entity of the world, and let $S$ be the set of all such sensorial experiences. This set will be defined under a given interpretation of the experimental environment available to the model, and can refer to an individual object of the world, or a category of objects, or a category of perceptual features. The set $S$ in turn belongs to the set of all classes of stimuli $\mathscr{S}$ available in the experiment. For a class $S \in \mathscr{S}$ we can define the two sets:

$$X_{S,i} = \left\{x_i(s_j) : s_j \in S\right\}; \quad \overline{X}_{S,i} = \left\{x_i(s_j) : s_j \in S' \neq S \in \mathscr{S}\right\}. \tag{4.16}$$

We can then associate with class $S$ a set of units in the map, by ranking it with the following function:

$$r(S,i) = \frac{\mu_{X_{S,i}} - \mu_{\overline{X_{S,i}}}}{\sqrt{\frac{\sigma_{X_{S,i}}}{|X_{S,i}|} + \frac{\sigma_{\overline{X_{S,i}}}}{|\overline{X}_{S,i}|}}}, \tag{4.17}$$

where $\mu$ is the average and $\sigma$ the standard deviation of the values in the two sets, and $|\cdot|$ is the cardinality of a set. Now the following relation can be established as the population code of a class $S$:

$$p(S) : \mathscr{S} \to \{\langle i_1, i_2, \cdots, i_M \rangle : r(S,i_1) > r(S,i_2) > \cdots > r(S,i_M)\}, \tag{4.18}$$

where $M$ is a given constant, typically one order of magnitude smaller than the number of units in the map. An alternative is to keep $M$ variable, and to derive the number of coding neurons by fixing a threshold on their ranking $r(S, i)$. The population code $p(S)$ computed with (4.18) can be used to classify a stimulus $s$ in an expected category:

$$c(s) = \arg\max_{S \in \mathcal{S}} \left\{ \sum_{j=1\cdots M} \alpha^j x_{p(S)_j}(s) \right\}, \tag{4.19}$$

where $p(S)_j$ denotes the $j$-th element in the ordered set $p(S)$, and $\alpha$ is a constant that is close, but smaller, than one.

In a typical experiment aimed at studying semantic representations, the set $\mathcal{S}$ of the environmental experiences available to the model is known in advance, and is partitioned a priori in meaningful categories $S_i$. The settlement of a semantic representation in a map of the model can be established by testing the discriminatory power of its population coding:

$$a(S) = \frac{|\{s : s \in S \wedge c(s) = S\}|}{|S|}. \tag{4.20}$$

It should be noted that the use of predefined categories, even if common in population code analysis, introduces a questionable assumption, known as the *labeling problem* (Lehky et al. 2013). An alternative is to derive an intrinsic code, that takes into consideration the relative activation of the neurons, avoiding to attach external labels to coding neurons, an example is the multidimensional scaling approach (Borg and Lingoes 1987; Borg and Groenen 2010). In most of the models that will be presented in the second part of the book the use of Eq. (4.20) is justified, because adopting external labels corresponds to the normativity of the language in naming things. Models exploring the emergence of pre-linguistic concepts will not use Eq. (4.20), and apply an internal interpretation of the codes instead, as in Sect. 6.1.4. If a consistent coding has been established, and if this coding can be assumed to be a genuine representation (see Sects. 3.1.2 and 3.1.3), Eq. (4.20) will also give the amount of potential *misrepresentation* of the model: the possibility that a stimulus $s_i$ will be taken as caused by an entity that is different from the category it actually belongs to.

Note that the analysis described so far does not introduce any alteration in the neural mechanisms of the model: the activation $x_i$ evaluated here is the result of equations of the kind (4.13), within maps developed using Eqs. (4.10), (4.11), and (4.12). The result of (4.18) is purely a statistical analysis applied to the output of a developed map, in order to establish the possible coding of external stimuli by a distributed population of units.

## References


Abbott, L. F., Rolls, E., & Tovee, M. J. (1996). Representational capacity of face coding in monkeys. *Cerebral Cortex, 6*, 498–505.

Ashby, W. R. (1947). Principles of the self-organizing dynamic system. *The Journal Of General Psychology, 37*, 125–128.

Ashby, W. R. (1962). Principles of the self-organizing system. In H. V. Foerster & G. W. Zopf (Eds.), *Principles of Self-Organization: Transactions of the University of Illinois Symposium* (pp. 255–278). New York: Pergamon.

Bednar, J. A. (2002). Learning to see: Genetic and environmental influences on visual development. PhD thesis, University of Texas at Austin, Tech report AI-TR-02-294.

Bednar, J. A. (2009). Topographica: Building and analyzing map-level simulations from Python, C/C++, MATLAB, NEST, or NEURON components. *Frontiers in Neuroinformatics, 3*, 8.

Bednar, J. A. (2014). Topographica. In D. Jaeger & R. Jung (Eds.), *Encyclopedia of computational neuroscience* (pp. 1–5). Berlin: Springer.

Belousov, B. (1959). Periodically acting reaction and its mechanism. *Collection of Abstracts on Radiation Medicine, 147*, 145. Originale in lingua russa.

Bénard, H. (1900). Les tourbillons cellulaires dans une nappe liquide. *Revue Générale des Sciences, 11*, 1261–1271, 1309–1328.

Borg, I., & Groenen, P. (2010). *Modern multidimensional scaling: Theory and applications* (2nd ed.). Berlin: Springer.

Borg, I., & Lingoes, J. (1987). *Multidimensional similarity structure analysis*. Berlin: Springer.

Bower, J. M., & Beeman, D. (1998). *The book of GENESIS: Exploring realistic neural models with the GEneral NEural SImulation System* (2nd ed.). New York: Springer

Bowers, J. (2009). On the biological plausibility of grandmother cells: Implications for neural network theories in psychology and neuroscience. *Psychological Review, 116*, 220–1078.

Broad, C. D. (1925). *The mind and its place in nature*. London: Kegan Paul.

Brunel, N., & Nadal, J. P. (1998). Mutual information, fisher information, and population coding. *Neural Computation, 10*, 1731–1757.

Carandini, M., & Heeger, D. (2012). Normalization as a canonical neural computation. *Nature Reviews Neuroscience, 13*, 51–62.

Cerreira-Perpiñán, M., & Goodhill, G. J. (2004). Influence of lateral connections on the structure of cortical maps. *Journal of Neurophysiology, 92*, 2947–295.

Chikazoe, J., Lee, D. H., Kriegeskort, N., & Anderson, A. K. (2014). Population coding of affect across stimuli, modalities and individuals. *Nature Neuroscience, 17*, 1114–1122.

Chirimuuta, M. (2014). Minimal models and canonical neural computations: The distinctness of computational explanation in neuroscience. *Synthese, 191*, 127–153.

Churchland, P. M. (1989). *A neurocomputational perspective: The nature of mind and the structure of science*. Cambridge: MIT.

Clayton, P., & Davies, P. (Eds.) (2006). *The re-emergence of emergence: The emergentist hypothesis from science to religion*. Oxford: Oxford University Press.

Cottrell, M., & Fort, J. (1987). Étude d'un processus d'auto-organisation. *Annales de l' institut Henri Poincaré, 23*, 1–20.

Dayan, P., & Abbott, L. F. (2001). *Theoretical neuroscience*. Cambridge: MIT.

de la Rocha, J., Doiron, B., Shea-Brown, E., Josić, K., & Reyes, A. (2007). Correlation between neural spike trains increases with firing rate. *Nature, 448*, 802–809.

Eliasmith, C., & Trujillo, O. (2014). The use and abuse of large-scale brain models. *Current Opinion in Neurobiology, 25*, 1–6.

Erwin, E., Obermayer, K., & Schulten, K. (1992a). Self-organizing maps: Ordering, convergence properties and energy functions. *Biological Cybernetics, 67*, 47–55.

Erwin, E., Obermayer, K., & Schulten, K. (1992b). Self-organizing maps: Stationary states, metastability and convergence rate. *Biological Cybernetics, 67*, 35–45.

Gerstner, W., & Kistler, W. M. (2002). Mathematical formulations of Hebbian learning. *Biological Cybernetics, 87*, 404–415.
Gilbert, C. D., Hirsch, J. A., Wiesel, T. N. (1990). Lateral interactions in visual cortex. *Cold Spring Harbor Symposia on Quantitative Biology, 55*, 663–677. Cold Spring Harbor Laboratory Press.
Grinvald, A., Lieke, E. E., Frostig, R. D., & Hildesheim, R. (1994). Cortical point-spread function and long-range lateral interactions revealed by real-time optical imaging of macaque monkey primary visual cortex. *Journal of Neuroscience, 14*, 2545–2568.
Gross, C. (2002). Genealogy of the "grandmother cell". *Neuroscience, 8*, 512–518.
Haken, H. (1978). *Synergetics – An introduction, nonequilibrium phase transitions and self-organization in physics, chemistry and biology* (2nd ed.). Berlin: Springer.
Hasker, W. (1999). *The emergent self.* Ithaca: Cornell University Press.
Hines, M., & Carnevale, N. (1997). The NEURON simulation environment. *Neural Computation, 9*, 1179–1209.
Hinton, G. E., McClelland, J. L., & Rumelhart, D. E. (1986). Distributed representations. In D. E. Rumelhart & J. L. McClelland (Eds.), *Parallel distributed processing: Explorations in the microstructure of cognition* (pp. 77–109). Cambridge: MIT.
Hou, C., Pettet, M. W., Sampath, V., Candy, T. R., & Norcia, A. M. (2003). Development of the spatial organization and dynamics of lateral interactions in the human visual system. *Journal of Neuroscience, 23*, 8630–8640.
Hubel, D., & Wiesel, T. (1962). Receptive fields, binocular interaction, and functional architecture in the cat's visual cortex. *Journal of Physiology, 160*, 106–154.
Hubel, D., & Wiesel, T. (1968). Receptive fields and functional architecture of monkey striate cortex. *Journal of Physiology, 195*, 215–243.
Hunt, J. J., Bosking, W. H., & Goodhill, G. J. (2011). Statistical structure of lateral connections in the primary visual cortex. *Neural Systems & Circuits, 1*, 1–12.
Kanerva, P. (1993). Sparse distributed memory and related models. In M. Hassoun (Ed.), *Associative neural memories: Theory and implementation*. Oxford: Oxford University Press.
Kauffman, S. A. (1993). *The origins of order – Self-organization and selection in evolution*. Oxford: Oxford University Press.
Kauffman, S. A. (2008). *Reinventing the sacred: A new view of science, reason, and religion*. New York: Basic Books.
Kim, J. (2006). Emergence: Core ideas and issues. *Synthese, 151*, 547–559.
Kohonen, T. (1982). Self-organizing formation of topologically correct feature maps. *Biological Cybernetics, 43*, 59–69.
Kohonen, T. (1984). *Self-organization and associative memory*. Berlin: Springer.
Kohonen, T. (1995). *Self-organizing maps*. Berlin: Springer.
Kriegeskorte, N. (2009). Relating population-code representations between man, monkey, and computational models. *Frontiers in Neuroscience, 3*, 363–373.
Lehky, S. R., Sereno, M. E., & Sereno, A. B. (2013). Population coding and the labeling problem: Extrinsic versus intrinsic representations. *Neural Computation, 25*, 2235–2264.
Lennie, P. (2003). The cost of cortical computation. *Current Biology, 13*, 493–497.
Linde, Y., Buzo, A., & Gray, R. (1980). An algorithm for vector quantizer design. *IEEE Transactions on Communications, 28*, 84–95.
Markram, H. (2006). The blue brain project. *Nature Reviews Neuroscience, 7*, 153–160.
Mastronarde, D. N. (1983). Correlated firing of retinal ganglion cells: I. Spontaneously active inputs in X- and Y-cells. *Journal of Neuroscience, 14*, 409–441.
Meyers, E. M., Freedman, D. J., Kreiman, G., Miller, E. K., & Poggio, T. (2008). Dynamic population coding of category information in inferior temporal and prefrontal cortex. *Journal of Neurophysiology, 100*, 1407–1419.
Miller, K. D., & MacKay, D. J. C. (1994). The role of constraints in Hebbian learning. *Neural Computation, 6*, 100–126.
Milner-Brown, H. S., Stein, R. B., & Yemm, R. (1973). Changes in firing rate of human motor units during linearly changing voluntary contractions. *Journal of Physiology, 230*, 371–390.

Olshausen, B. A., & Field, D. J. (1996). Natural image statistics and efficient coding. *Network: Computation in Neural Systems, 7*, 333–339.
Olshausen, B. A., & Field, D. J. (2004). Sparse coding of sensory inputs. *Current Opinion in Neurobiology, 14*, 481–487.
Pasupathy, A., & Connor, C. E. (2002). Population coding of shape in area v4. *Nature Neuroscience, 5*, 1332–1338.
Plebe, A. (2001). Self-organizing map approaches to the traveling salesman problem. In M. Maggini (Ed.), *Limitations and Future Trends in Neural Computation, NATO Advanced Research Workshop*, 22–24 Oct 2001, Siena.
Plebe, A., & Anile, M. (2001). A neural-network-based approach to the double traveling salesman problem. *Neural Computation, 14*(2), 437–471.
Pribram, K. H. (1971). *Languages of the brain: Experimental paradoxes and principles in neuropsychology*. Englewood Cliffs: Prentice Hall.
Prigogine, I. (1961). *Introduction to thermodynamics of irreversible processes*. New York: Interscience.
Quian Quiroga, R., & Kreiman, G. (2010). Measuring sparseness in the brain: Comment on bowers (2009). *Psychological Review, 117*, 291–297.
Quian Quiroga, R., & Panzeri, S. (Eds.) (2013). *Principles of neural coding*. Boca Raton: CRC.
Quian Quiroga, R., Reddy, L., Koch, C., & Fried, I. (2007). Decoding visual inputs from multiple neurons in the human temporal lobe. *Journal of Neurophysiology, 4*, 1997–2007.
Quian Quiroga, R., Kreiman, G., Koch, C., & Fried, I. (2008). Sparse but not 'grandmother-cell' coding in the medial temporal lobe. *Trends in Cognitive Sciences, 12*, 87–91.
Ritter, H., Martinetz, T., & Schulten, K. (1992). *Neural computation and self-organizing maps*. Reading: Addison Wesley.
Rolls, E., & Tovee, M. J. (1995). Sparseness of the neuronal representation of stimuli in the primate temporal visual cortex. *Journal of Neurophysiology, 73*, 713–726.
Ruelle, D., & Takens, F. (1971). On the nature of turbulence. *Communications in Mathematical Physics, 20*, 167–192.
Rumelhart, D. E., & McClelland, J. L. (1986). On learning the past tenses of English verbs. In D. E. Rumelhart, & McClelland, J. L. (Eds.) *Parallel distributed processing: Explorations in the microstructure of cognition* (pp. 216–271). Cambridge: MIT.
Sakai, K., Naya, Y., & Miyashita, Y. (1994). Neuronal tuning and associative mechanisms in form representation. *Learning and Menory, 1*, 83–105.
Singer, W. (1995). Synchronization of neuronal responses as a putative binding mechanism. In *The handbook of brain theory and neural networks*. Cambridge: MIT.
Sirosh, J., & Miikkulainen, R. (1997). Topographic receptive fields and patterned lateral interaction in a self-organizing model of the primary visual cortex. *Neural Computation, 9*, 577–594.
Sirosh, J., Miikkulainen, R., & Choe, Y. (Eds.) (1996). *Lateral interactions in the cortex: Structure and function*. Austin: The UTCS Neural Networks Research Group.
Stettler, D. D., Das, A., Bennett, J., & Gilbert, C. D. (2002). Lateral connectivity and contextual interactions in macaque primary visual cortex. *Neuron, 36*, 739–750.
Stevens, J. L. R., Law, J. S., Antolik, J., & Bednar, J. A. (2013). Mechanisms for stable, robust, and adaptive development of orientation maps in the primary visual cortex. *JNS, 33*, 15,747–15,766.
Turrigiano, G. G., & Nelson, S. B. (2004). Homeostatic plasticity in the developing nervous system. *Nature Reviews Neuroscience, 391*, 892–896.
von der Malsburg, C. (1973). Self-organization of orientation sensitive cells in the striate cortex. *Kybernetic, 14*, 85–100.
von der Malsburg, C. (1995a). Binding in models of perception and brain function. *Current Opinion in Neurobiology, 5*, 520–526.
von der Malsburg, C. (1995b). Network self-organization in the ontogenesis of the mammalian visual system. In S. F. Zornetzer, J. Davis, C. Lau, & T. McKenna (Eds.), *An introduction to neural and electronic networks* (2nd ed., pp. 447–462). New York: Academic.

Willshaw, D. (2006). Self-organization in the nervous system. In R. Morris & L. Tarassenko (Eds.), *Cognitive systems: Information processing meets brain science* (pp. 5–33). Amsterdam: Elsevier.

Willshaw, D. J., & von der Malsburg, C. (1976). How patterned neural connections can be set up by self-organization. *Proceedings of the Royal Society of London, B194*, 431–445.

Yu, H., Farley, B. J., Jin, D. Z., & Sur, M. (2005). The coordinated mapping of visual space and response features in visual cortex. *Neuron, 47*, 267–280.

Zhabotinsky, A. (1964). Periodical process of oxidation of malonic acid solution. *Biophysics, 56*, 178–194.

# Part II
# Meaning from Neurons

This second part of the book ventures into the core of neurosemantics: the project of explaining linguistic meaning in terms of computations done by the brain. It is an enterprise at the edge of the available state-of-the-art knowledge in neuroscience, and the understanding of brain computational mechanisms, highlighted in the first part, undoubtedly audacious, and still in its early infancy. However, we conceive neurosemantics as the natural evolution of a long standing project, started in the early days of Boole's logic: the idea that semantics can be construed and explained in mathematical terms.

The first chapter of this part traces this evolution, highlighting the connections as well as the discontinuities. The first one is the cognitive turn that brought in the mind, excluded by classical formal semantics. The second divergence is from the weak mathematical framework of cognitive semantics, stuck within a much too abstract concept of computation, towards a sound mathematical foundation, empirically grounded in how the brain computes: neurosemantics.

The next three chapters present samples of neurosemantics in practice. All the premises of the first part are taken together, and turned into the implementation of a series of models, seeking to capture aspects of linguistic meaning at the neurocomputational level. A unified approach, based essentially on the algorithmic principles described in Chap. 4, is applied to simulating various areas of the cortex, involved in three aspects of language understanding: the early building of names for visual objects, the class of color terms and their use as adjectives, and the semantics of moral terms. Needless to say, even if these three models explore important semantic phenomena, their coverage with respect to the complexity of real languages is modest indeed. Their main aim is exploratory, demonstrations of how the vertiginous cliff, from the level of a neural signal up to linguistic meaning, can be climbed, even if with more than a few missing details, by a brain friendly mathematics.

The last chapter is still a demonstration of neurosemantics, but this time, covering a selection of models developed by other research groups, that share a view largely compatible with this book, and that we feel justified in classifying as neurosemantics. These models target aspects of language meaning not explored in our own previously mentioned models, such as word order, compositionality, and the semantics of numbers.

In conclusion, we should once again remind the reader, that currently the overall picture on how the brain captures linguistic semantics is sketchy and largely deficient. Neurosemantics is a research effort that would not have been possible a few decades ago. Today, we see it as the most appropriate effort in explaining language, and in this book, try to outline its domain, to describe its methodology, and to portray its first steps.

# Chapter 5
# Semantic Theories

**Abstract** Semantics, in the sense used in this book, seeks to understand the meaning of words and sentences, explaining the relations between expressions in a natural language and the world. This chapter intends to give a short account on semantics as has been developed before neurosemantics, and trace the path that naturally lead to its neuro form. The link that, in our view, connects the milestones of semantic theories to neurosemantics, is the aim of constructing precise mathematical models of the relations between linguistic entities and their referents. Therefore, much attention will be given to the transition from a descriptive semantics to its mathematical foundation in modern logic. It will be argued that the unsatisfactory aspect of this project was to have neglected the mind, which became on the contrary, the main object of investigation during the cognitive turn. We will describe the vicissitudes of cognitive semantics, its merits, and the counterside of a serious weakness in the level and robustness of its mathematical modeling.

## 5.1 Logic and Meaning

The study of linguistic meaning is not a new endeavor, having interested philosophers from antiquity, but philosophers of the early part of the twentieth century made it one of their central areas of inquiry. This was the result of the interaction of a number of disciplines in ferment at the time, one being the renewed interest in the formal study of reasoning, known as logic, in the pursuit of finding an epistemological foundation for mathematics. Natural language because of its fundamental link to human reasoning, thus also became an area of mathematical investigation, with mathematical logic having a pronounced influence on the study of semantics in the years that followed. This section will briefly review those proposals, that in the not so distant past, were put forth to explain linguistic meaning in mathematical and logical terms, and in some respects are on the same thread of the story leading to neurosemantics (Plebe 2004).

 
A. Plebe, V.M. De La Cruz, *Neurosemantics*,
Studies in Brain and Mind 10, DOI 10.1007/978-3-319-28552-8_5

### 5.1.1 The Mathematics of Thinking

To conceive thinking in mathematical terms is not really new at all, especially for what concerns rational argumentation. Gottfried Leibniz suggested that a specific kind of calculus would be the key to settling all human conflicts and disagreements, in his words:

> Quo facto, quando orientur controversiae, non magis disputatione opus erit inter duos philosophos, quam inter duos computistas. Sufficiet enim calamos in manus sumere sedereque ad abacos, et sibi mutuo dicere: calculemus! Leibniz (1684)

For this exhortation to be feasible a new mathematics was necessary, a *calculus ratiocinator* planned by Leibniz as the last and major effort of his life, but was never commenced. His desired mathematics of thinking was conceived as an external tool to accomplish any type of reasoning, and did not explicitly entail that our mental way of reasoning was mathematical in its essence.

It was, on the contrary, the assumption of Thomas Hobbes, who not much later, asserted:

> For reason [...] is nothing but reckoning (that is, adding and subtracting) of the consequences of general names agreed upon for the marking and signifying of our thoughts; I say marking them, when we reckon by ourselves; and signifying, when we demonstrate or approve our reckonings to other men. (Hobbes 1651, Cap V)

Hobbes did not attempt to elaborate the mathematics behind reasoning either, it was neither among his intentions nor his possibilities, considering he did not master mathematics like Leibniz did. Only about two centuries later was the first mathematics of reasoning laid down, by George Boole (1854b). His great effort was in using standard algebra, giving variables and operations a new meaning, related to mental thinking.

Variables, for which Boole used the higher alphabetic letters, such as $x, y, z, \ldots$, which denote sets of objects satisfying a specific concept, for example $x$ might be the set of `animal` and $y$ the set of `green` entities. The product operation, whose correspondent symbol is omitted as in ordinary algebra, corresponds to the intersection set operation $\cap$, so that $xy$, in our example, is the set of green animals like frogs and lizards. The $+$ operator corresponds to the union $\cup$. There are two possible constant values: 1 corresponding to the universe of objects, and 0 to the empty set, therefore $1 - x$ is the set of all non animated objects.

The basic operations are ruled by the following set of basic laws:

$$xy = yx \tag{5.1}$$

$$x + y = y + x \tag{5.2}$$

$$z(x + y) = zx + zy \tag{5.3}$$

$$x = y \Rightarrow zx = zy \tag{5.4}$$

$$x = y \Rightarrow z + x = z + y \tag{5.5}$$

$$x^2 = x \tag{5.6}$$

$$x + x = x \tag{5.7}$$

where $=$ is the identity symbol. Equations (5.1) and (5.2) are commutative properties, Eq. (5.3) is the associative, and (5.4), (5.5) are identity properties. Equation (5.6) is called the dual property, and it is at the core of the deductive system proposed by Boole. For example, from (5.6) derives that $x(1 - x) = 0$, as in our example, that nothing can be an animal and not an animal at the same time.

Boole went further in relating algebraic expressions to propositions of natural language, for this purpose he introduced a special variable, $v$, the indefinite class. This was in fact Boole's expedient to express *quantification*, which found a much more elegant solution in Frege (see Sect. 5.1.2). His three "primary" propositions are those listed in Table 5.1, where $f_S$ is an arbitrary expression denoting a subject, and $f_P$ is an arbitrary algebraic predicative expression.

Note in particular propositions the use of $v$ as surrogate for quantification. The logic system of Boole is completed with an elaborate methodology for actually "solving" systems of equations, corresponding to propositions. It is divided into three main phases: elimination, reduction, and development. The first two are direct extensions of the ordinary methods of algebraic manipulations, such as identification of superfluous variables and their elimination, and the reduction of a system of equations to the minimum number. What Boole calls "development" is instead specific to the meaning of the symbols in his logical system.

**Table 5.1** The three primary propositions in Boole's algebra. The second column from the left is the general format of the proposition, the third column is an example of the proposition, with its algebraic translation in the rightmost column

| | | | |
|---|---|---|---|
| Universal | $f_S = f_P$ | Computer scientists are animals with keyboards not so keen on gym | $p = ab(1 - s)$ |
| Particular predicate | $f_S = vf_P$ | Computer scientists wearing glasses are nearsighted | $op = vm$ |
| Particular subject and predicate | $vf_S = vf_P$ | Some computer scientists with age become philosophers | $vtp = vf$ |
| Symbols used in the examples: | | $a =$ animals | |
| | | $b =$ with keyboard | |
| | | $f =$ philosopher | |
| | | $m =$ nearsighted | |
| | | $o =$ with glasses | |
| | | $p =$ computer scientists | |
| | | $s =$ gym addict | |
| | | $t =$ aged | |
| | | $v =$ the indefinite class | |

Analyzing this system in depth is not the goal of this book, but the remarkable fact we would like to highlight is that one of the main objectives Boole had in inventing his system was to describe the mental processes of reasoning. This is the aspect of his work that was completely removed in the ensuing developments of logic. Bertrand Russell (1918) in declaring his admiration for the pioneering work of Boole, alluded to his extravagance in connecting logic and mind: "Pure Mathematics was discovered by Boole in a work he called *The Laws of Thought*".

### 5.1.2 The Mathematics of Meaning

Boole opened the road to logic, but his system was constrained in making use of symbols and tools inherited from ordinary algebraic calculus, which was invented for working with numbers, and despite his great efforts, was unmalleable for new purposes. Contemporary logic is mainly due to Gottlob Frege, and one of his first ideas was to invent from scratch a way of formally expressing "concept". In fact, he named his system "Begriffsschrift" (Frege 1879). We can anticipate right away to the interested reader that, unfortunately, no one has ever used the Begriffsschrift after Frege invented it, but this nonetheless, some of the constituents of his system became the foundation of contemporary logic.

The Begriffsschrift, in its neglected aspect, was a curious and ambitious new way of writing that breaks the common sequence from left to right and from top to bottom, of western languages (and mathematical writing too). It develops in two dimensions. The basic form is the assertion: $\vdash$——$A$, $A$ is true. Unlike in Boole's system, variables like $A$ are now propositions, sentences which can be considered as true or false. The two basic connectives are the negation and the implication, drawn as in following, with the current notation below:

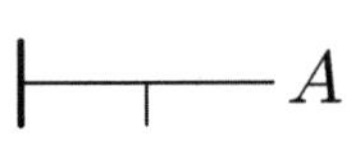

$\neg A$

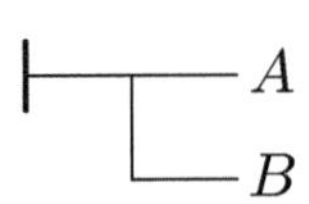

$B \supset A$

in every logical system, it is possible to define axiomatically two connectives only, and all the remaining can be derived, here for example, are the conjunction and the disjunction in the pictorial Begriffsschrift form:

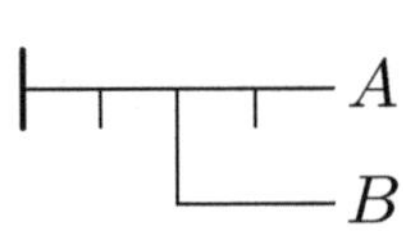

$A \wedge B$

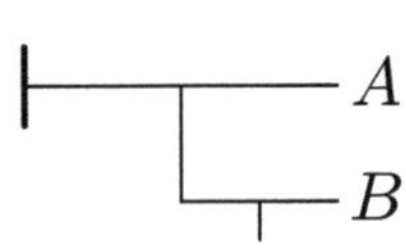

$B \vee A$

The breakthrough ofFrege's system is in two radical innovations with respect to Boole: the function and the quantification. He did not simply borrow the idea of function from calculus, as a mathematician he first rigorously scrutinized the concept and the use of "function" in ordinary mathematics (Frege 1904), and then used for semantics his purified rigorous account of function, as an entity that needed to be saturated, much like chemical ions. Only when saturated by its argument does the function become a real object (Dummett 1973). Function and argument in semantics correspond to any possible decomposition of a simple proposition into two components. Unlike in calculus, functions return only truth values.

Frege's project was ambitious, he wanted to base arithmetic upon logic, and in addition to the Begriffsschrift he developed a full axiomatic system (Frege 1884). It is well known that this system was corrupted by the excess of freedom in defining functions, as discovered by Bertrand Russell, who posed the famous question:

> You state that a function, too, can act as the indeterminate element. This I formerly believed, but now this view seems doubtful to me because of the following contradiction. Let *w* be the predicate: to be the predicate that cannot predicate of itself. Can *w* be predicated of itself? From each answer its opposite follows. Therefore we must conclude that *w* is not a predicate. Likewise there is no class (as a totality) of those classes which, each taken as a totality, do not belong to themselves. From this I conclude that under certain circumstances a definable collection does not form a totality.
>
> (Russell 1902, pp. 124–125)

The failure of his project broke Frege's heart, but his inventions inspired the work of Russell himself, the first work of Ludwig Wittgenstein (1922b), including the equivalence between meaning and truth conditions, that of Rudolph Carnap (1928) and dominated the ensuing developments in logic (van Benthem and ter Meulen 1996). It is not useful for this book to delve further into contemporary logic, but there are just a couple of aspects of Frege's work that are worth mentioning.

On one side, he has been heralded as the most fierce defender of anti-psychologism in logic, representing therefore, a drastic discontinuity from Boole, in purging everything pertaining to the mental sphere from the abstract elucidation of semantics in formalized logical terms (Baker and Hacker 1989). In criticizing Husserl's philosophy of arithmetic, Frege (1894) concludes by saying: "In reading this work, I was able to gauge the devastation caused by the influx of psychology into logic; and I have here made it my task to present this damage in a clear light. The mistakes which I thought it my duty to show reflect less upon the author than they are the result of a widespread philosophical disease." Despite such inveighing, the celebration of Frege as an apologist of the purity of logic against the contamination from investigations on the mind, in antagonism with Boole, is perhaps a picture that has been amplified within the analytic philosophy of the mid 1900s, and according to Vassallo (2000), there is much more Boole and Frege have in common, from the standpoint of psychologism, than meets the eye.

In one of the few writings of Frege (1892) exclusively focused on the semantics of natural language, a new element is suggested, which in our opinion, is a valid example of the difficulties of a semantics segregated from the mind. The standard meaning of singular terms should be in their contribution to the truth conditions in the proposition in which they appear (Russell 1905). According to Frege, there is something else, which he called *Sinn*, usually translated as "sense". The best way to get convinced of the *Sinn* component is by comparing co-referential expressions, i.e. referring to exactly the same object in the world, for example:

1. `the best mathematician of the year in 1967 at Michigan University and lecturer of Analysis at Berkeley University in 1968;`
2. `the serial murderer who killed 3 people and injured 23 others between 1978 and 1995;`

are two different ways of denoting the same person: Theodore Kaczynski, infamously known as *Unabomber*. Even if the reference is the same, expressions found in 1 convey information on his high intellectual achievement and a service given to the university community. Expression 2, on the contrary, includes sinister information about his behavior. The two expressions, even if referring to the same entity, challenge the logical substitutivity principle, for example while equating expression 1 with itself is trivially true, equating 1 with 2 is informatively true. Even worse, substitution in belief-contexts may change the truth values: for an unaware and diligent student of Kaczynski to believe 1 was true, but certainly not to believe 2.

This is due, according to Frege, to the difference in "mode of presentations" of the referent, that is what the *Sinn* component of the expressions is. Its precise nature was not detailed by Frege, and remained a struggle for generations of philosophers (Dummett 1973; Peacocke 1992; Biro and Kotatko 1995). The point is that according to Frege, and the Fregean tradition, all that can explain *Sinn* should withhold any reference to the mind, and should be in the elements of the expressions whether or not anyone ever believed it. As remarked by Margolis and Laurence (2007, p. 20):

> [...] the sense-based solution to the mode of presentation problem says that the reference of a word or internal representation is mediated by a sense that we grasp. But what exactly does grasping consist in? Clearly, grasping is a metaphor for a cognitive relation that needs to be explicated.

It is hard to see how *Sinn* can ever be explained without a mental account.

### 5.1.3 Logic in the Brain?

During the most flourishing period of logic, when tainting its study with investigations on the mind was heresy, an even more extreme affair was proposed, to connect logic with brain circuits. This bold enterprise was carried out by two scientists with

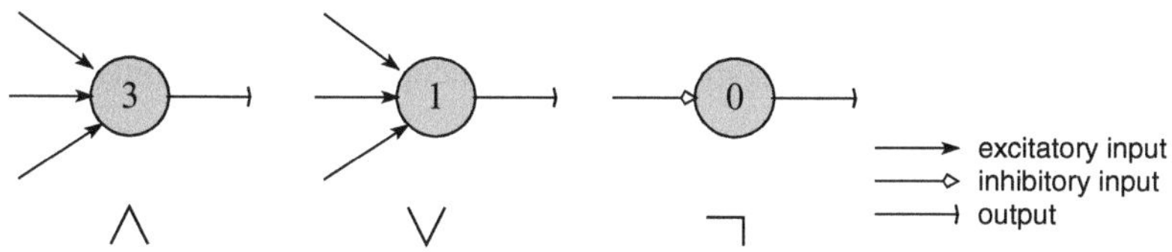


**Fig. 5.1** Neurons in the model of McCulloch and Pitts: on the *left* the implementation of logical conjunction, in the *middle* the disjunction, on the *right* the negation operation. The number inside the neuron is its threshold

complementary competences, on one side Warren McCulloch, neurophysiologist and psychiatrist, and on the other, Walter Pitts who studied logic with Rudolf Carnap. They believed that, despite the apparent entanglement of neural signals, due to the huge difficulty in implementing computations in organic matter, the brain performs crystalline logical operations (McCulloch and Pitts 1943). The idea was tossed around in the group with which McCulloch and Pitts worked, led by Nicolas Rashevsky (1938), a pioneer in the use of mathematical tools in biology. He was convinced that the best way to abstract mathematically the behavior of the brain was in assuming binary values for the neurons, furthermore, he sketched a possible scheme of logical exclusive disjunction based on summation and subtraction of signals.

McCulloch and Pitts continued the effort, completing a theory of logic based on neurons, adapting Carnap's formalism in a rigorous way. There are two types of synaptic connections only: excitatory and inhibitory, and two types of signals, corresponding to the logical truth values. An intrinsic feature of all neurons is a threshold, corresponding to the net number of `true` values necessary to produce `true` as output value. The neuron on the left in Fig. 5.1, implements the logical conjunction of three inputs, and is `true` only when all three inputs are `true`, for the neuron in the center one single `true` input is enough, therefore working as a disjunction logical connective. In addition, neurons can have recurrent connections, realizing memories of logical states. The neural system, in this abstract model, has the same semantic power of a logical system: it is able to represent the meaning of any linguistic proposition in terms of truth conditions.

The brain as a logic machine was a fascinating hypothesis, and galvanized more than one scholar at that time, from the doctoral thesis of Minsky (1954), to the finite state automata of Kleene (1956), and the parallel brain-computer made by von Neumann (1958). Quite soon, Hodgkin and Huxley (1952) began to disclose a very different picture of how neurons behave, and today we know well that McCulloch and Pitts' idea was simply wrong (see Sect. 2.1). They too became well aware of the different direction the growing neuroscientific evidence was pointing to, while they were working on the visual system. Already in their paper on the perception of visual forms in the cortex (Pitts and McCulloch 1947), they used mathematical formulations that were far from logic (group invariant operators), and in the work on frog vision the logic approach was completely abandoned (Lettvin et al. 1959). Nevertheless, the attempts made by McCulloch and Pitts are rich in historical merit

(Piccinini 2004), in that they launched the idea that the brain performs computations in the precise sense of Turing computations, a concept that is commonly held today by scholars in the field.

Their attempts endorse the pertinence of starting the second part of this book with classic logic. Even if the founders of the first mathematics of semantics fiercely fought against any relationship with the mental mechanisms of semantics, there have been scholars that have believed that logic not only should describe something that goes on in the mind, but that it is exactly how the brain works. Starting however, from an existing and well defined mathematical framework and expecting the brain to work that way, is a venturesome move. It is thus, better to leave logic aside, and gradually move towards a more mind friendly semantics.

## 5.2 Semantics Meets the Mind

After the cognitive shift in the 1970s, cognitive semantics along with cognitive approaches to grammar, emerged as one of the main branches of cognitive linguistics, which concentrates on investigating the relationship between language, the mind and socio-physical experience. Cognitive linguistics, described as a "movement", by one of its historic proponents, Ronald Langacker (1987), is characterized by a set of core commitments and guiding principles, which in turn have produced a number of complimentary, different, overlapping and often competing theories. This section will delineate the historical development of the field of cognitive semantics, and its efforts in taking the study of the making of meaning, once again, within the context of the mind.

### *5.2.1 The Unfulfilled Promise*

It is not possible to talk about explaining natural language in computational terms, without mentioning the extensive work of Noam Chomsky and his school, even if semantics was exactly one of their neglected aspects of language, in favor of syntax, the way words are placed in order in sentences and the rules by which this is done. One of the greatest initial merits of Chomsky (1956, 1957, 1958) was that of coming up with a framework for a mathematical account of grammar, overcoming the limitations of the linguistic approaches that had been employed up until the 1950s, especially structuralism and behaviorism. This new approach provided a way to apply the intuitions developed by Harris (1951) in linguistics, in connection with the mathematical theory of formal languages, started by Thue (1906, 1912) and refined by Post (1921, 1947). At the core of the mathematical description of language there is the definition of abstract grammar:

$$G \stackrel{\text{def}}{=} \langle V_N, V_T, S, R \rangle \tag{5.8}$$

which elements are:

$$\begin{array}{ll} V_N & \text{set of non terminal elements} \\ V_T & \text{set of terminal elements, with } V_N \cap V_T = \emptyset \\ S & \text{root element, with } S \in V_N \\ R & \text{set of rewriting rules, each with the format } \langle P, Q \rangle \text{ with} \\ & P, Q \in (V_N \cup V_T)^* \text{ and } P \text{ with at least on element} \in V_N \end{array} \tag{5.9}$$

A language is defined as follows:

$$L(G) = \left\{ P \mid S \overset{*}{\underset{G}{\Longrightarrow}} P \right\} \cap V_T^* \tag{5.10}$$

where the operation $\overset{*}{\underset{G}{\Longrightarrow}}$ is the application of any rule available in $G$ an arbitrary number of times. Equation (5.10) tells that a language is the set of all sentences generated according to a grammar of the vocabulary of terminal elements $V_T$.

This abstract construct was just the basis for building a set of rules corresponding to the extant combination of words in the sentences of natural languages, and one of the most serious challenges in this enterprise was the definition of a Universal Grammar able to explain the great variety of syntactic features in the different languages of the world. This effort has led to a long evolution of the theory itself, reflected in a sequence of labels. In the Government and Binding theory (Chomsky 1981) the focus was on the relations between two words, not necessarily adjacent in the sentence, and the binding with anaphoric elements like pronouns. It was superseded by the Principles and Parameters (Chomsky and Lasnik 1993) theory, in which a small number of syntactic constructs, such as the $X$-bar, are held to be universal, with "parameters" that adapt their precise format to each language of the world. Throughout this evolution there has been a progressive relinquishing of the stringent mathematical formalization, which has become a barely perceivable background in the most recent elaboration of the theory, the Minimalist Program (Chomsky 1993, 1995), where the only syntactic operation left is `Merge`, roughly working as a rewriting rule.

Universal Grammar has produced nearly as many successful formal descriptions of syntactic phenomena, as logic did for semantic constructions. Both gained their powerful descriptive adequacy by using a mathematical formalism that is totally unrelated with mental processes. The difference is that while logicians defended their discarding the mental realm as a theoretical position, Chomsky and his school, on the contrary have put high claims on their mathematics as corresponding to mental processes of language understanding, and even brain processes. Katz (1981) was one of the first to call attention to the issue, in ontological terms. The objects of analysis in linguistics appear to be abstract entities, and especially in the generative grammar framework, close to mathematical objects. Yet Chomsky insisted that they were psychological entities. He did not deny the ambiguity (Chomsky and Halle 1968): "We use the term 'grammar' with a systematic ambiguity. On the one hand, the term refers to the explicit theory constructed by the linguist and proposed as a description of the speaker's competence. On the other hand, it refers to this competence itself." The fallacy, in the words of (Katz and Postal 1991, p.527), is that:

> We may assume that there is a domain of fact, A, instantiated by (1)–(6) [syntactic rules], studied in field A' and a domain of fact, B, concerned with human linguistic knowledge, its development, the biological structures [...] which determine it, etc., studied in field B'. Evidently, both domains A and B and fields A' and B' are characterized a priori in distinct ways. While A and B could turn out to be identical, they could also turn out to be distinct. Therefore, they cannot simply be *assumed* to be identical.

We would like to point out that the contradiction is even more puzzling, in that Chomskian linguistics departed from psychology programmatically, in disregarding the connection between syntax and meaning.

Katz characterized the ontological contradiction in Chomskian linguistics as the clashing between realism (in the sense of logical and mathematical realism) and conceptualism. Other dichotomies have been used by other authors to describe the symptoms of the defective Chomskian ontology. Seuren (2004a) uses "realism" in the opposite sense of Katz, as a theory that "aims at describing and specifying the workings of the hidden underlying reality that is the object of the theory under some formula of interpretation." It is the dual with "instrumentalism", a theory that "merely tries to capture observed regularities in terms of an algorithmically organized system or formal theory, without any causality or reality claim regarding the theoretical terms employed in the theory." Generative grammar is clearly a purely instrumentalist theory, yet their proposers affirm to be committed to realism, in Seuren's sense. The ontological contradiction, as put forth by Katz and Seuren, may be overcome by noting that, in principle, abstract computational or mathematical objects can be adequate for describing implementation as well, thanks to a mapping with concrete mechanisms of the implementation (see Sect. 3.1.5). However, the point is that this kind of mapping is not of concern within Chomskian linguistics, and this leads us to the epistemological aspect of the contradiction.

It has been focused on by (Stich and Ravenscroft 1994, p.15), who marked a distinction between "external" and "internal" inquires in the generative grammar project. A grammar, on the external view, "is nothing more than a useful systematization or axiomitation of linguistic intuitions." On the internal view, instead, "a grammar should not only capture (or entail) most linguistic intuitions, it should also be part of the mechanism that is causally responsible for the production of those intuitions, and for a variety of other linguistic capacities." The incoherence is that the focus in the development of generative grammar is in the descriptive adequacy, the simplicity, and the formal elegance of the theory, with no attention to the mental processes involved in language. Still, UG theorists have higher aspirations for grammar, that of magically transforming it into an internal account. But a grammar constructed on a strict external project could very well have principles that are quite at odds with anything that is subserved by a specific mental mechanism.

The gap between the polar ontology and epistemology in linguistics has even worsened throughout the years, as the aspirations for generative grammar have grown even higher, from mentalism and psychologism up to the physical level of brain and biology, in what is called Chomsky's "biolinguistic" view (Chomsky 2000, 5):

> It [the biolinguistic approach] is concerned with mental aspects of the world, which stand alongside its mechanical, chemical, optical and other aspects. It undertakes to study a real object in the natural world – the brain, its states, and its functions.

An example of this inner incoherence is the position with respect to one of the basic guiding principles in biology: Darwinian evolution. For Chomsky (1988, p.183) this principle in the case of language would be "a complete waste of time, because language is based on an entirely different principle than any animal communication system". The solution is the idea described by Botha (1999) as the "Fable of Instantaneous Language Evolution". In facing the impossibility to harmonize the abstraction of generative grammar with the facts of biological evolution, Chomsky (2005, p.11–12): insists on the antiscientific suggestion of an apparent miracle for the origin of language:

> With Merge available, we instantly have an unbounded system of hierarchically structured expressions. The simplest account of the "Great Leap Forward" in the evolution of humans would be that the brain was rewired, perhaps by some slight mutation, to provide the operation Merge, at once laying a core part of the basis for what is found at that dramatic "moment" of human evolution.

Another unproductive consequence of the ambiguity between the ontological status of generative grammar and neurophysiological reality is the postulate of a "language organ", often called the "language faculty". This entity is the essence of the ontological ambiguity, thought at the same time as a set of formal linguistic principles, and as a piece of the brain, tying in nicely with modularism in the sense of Fodor (1983). According to Chomsky (1995, p.167): "The human brain provides an array of capacities that enter into the use and understanding of language (the *language faculty*)". We already discussed in Sect. 3.3.3 the serious divergences between modularity as conceived by Fodor, but also by Carruthers, and what is known today about the organization of the brain. In the case of language as well, the hypothesis of a domain-specific, innate module corresponding to Chomsky's language faculty has not garnered substantial support from neurobiology. Certainly many questions regarding language processing in the brain remain open, but at this stage, the Chomskian picture looks less promising than that of a contribution of diverse brain regions, including non-specific areas (Stowe et al. 2004; Osterhout et al. 2007; Prat et al. 2007; Proverbio et al. 2009; Pulvermüller 2010).

In some sense, expecting a correspondence in the brain of the mathematical constructs of generative grammar, is similar to the attempt made by McCulloch and Pitts to adapt formal logic to the workings of neurons, but their idea was a provisional hypothesis, open to being falsified by empirical investigations, which they themselves were engaged in.

This brief account on generative grammar is not just the narration of an unfulfilled promise for a mathematical formalization of language meeting the mind. This unfulfilled promise, so to speak, has also been historically important in a number of other ways. The growing dissatisfaction of scholars working within the Chomskian project, encouraged several of them to move in a totally different scientific direction.

### 5.2.2 Cognitive Semantics

Syntactocentricsm was essential to Chomsky's early success, making the algorithmic formalization of linguistic rules relatively easier, it also undermined the mental reality of its foundation. During the rise of the brand new cognitive science, around 1960, a central idea was to imagine the mind as a computational device, subject to the theoretical laws discovered by Turing (1936) and Post (1947). The rewriting computations postulated by linguists for the grammars of natural languages would obviously be regarded as instances of the kind of mental algorithms envisaged by psychologists, but this natural merge did not happen, because isolating syntax required leaving meaning out of the picture, and therefore, most of the mind. In the same period an alternative project, under the name of "generative semantics", made serious and promising advances in rejoining cognitive science. Some linguists of the group lead by Chomsky, such as Postal, Lakoff, and McCawley, worked on combining syntactic structures with semantic logical representations. Chomsky was displeased and mounted a campaign against generative semantics, popularized as the "linguistics war" (Harris 1993), causing its rapid disappearance.

But the dissatisfaction did not disappear. Cognitive science was growing, offering simple arguments to overcome the theoretical reasons to reject mental analysis in traditional semantics. One of the most powerful was the argument against a "private language" made famous in philosophy by Wittgenstein (1953). In order for language to be understandable, it must be public, therefore, all that counts in its way of working cannot just be in the head of a speaker. Surely, the private language argument has had plenty of different interpretations, as well as consequences (Kripke 1982), like many passages in the *Philosophische Untersuchung*, our interpretation corresponds to the way this argument is typically used as a precaution against projects of mental semantics. Cognitive science can deflate the argument, finding in the external world, and in perceptual experience, the common ground for communication. Each individual mind develops an understanding of language based on its experience of a shared world, and it is enough to warrant reciprocal understanding.

A different reason for antimentalism had an epistemological basis, and was rejected by several scholars of logic positivism, such as Hempel and Carnap, and proponents of behaviorism (Smith 1986). Even if Wittgenstein never admitted affinities, his repeated plea for the observation of external behavior as the only source of knowledge, such as that found in his *linguistic games*, are not far from behaviorism. To confine the study of human phenomena to observable events has the worthy benefit of scientific rigor, but at the price of ending the investigation prematurely. The reaction of cognitive science was to reverse the object of research, putting the mind at the center. In the case of language, the reaction of cognitive science manifested itself in a number of approaches, all alternatives to both Chomskian grammar and logic semantics. They emerged in the 1970s and increasingly in the 1980s, collected under the label of "cognitive linguistics" (Croft and Cruse 2004; Geeraerts 2006; Brdar et al. 2011).

One of the main directions came from the critique of the autonomy of syntax. Lakoff (1986) was persuaded to abandon the generative grammar enterprise, becoming one of the leading exponents of cognitive linguistics, trying hard to construct rules for separating syntax from semantics, specifically in the *coordinate structure constraint*. It is one of the constraints on transformations that allows elements in phrase structures to be rearranged. For example, the *wh*-movement might allow a transformation like this:

```
Bob gave a book to Bill
what did Bob give to Bill?
```

but not when an item is coordinated with others, as in this case:

```
Bob gave a record and a book to Bill
*what did Bob give a record to Bill?
```

The coordinate structure constraint is purely syntactic, it mentions coordinate constituents and movement rules. However, Lakoff (1986) and Goldsmith (1985) found many counter examples, like:

```
Bob can drink two beers and still stay sober
How much can Bob drink and still stay sober?
```

Lakoff than went on to offer a semantic description of the conditions for moving coordinate conjuncts. In this example, the principle is that the constraint can be violated when the two conjuncts are kinds of natural sequences of events, causing, enabling or not preventing. Drinking that much does not prevent Bob from staying sober.

Thirty years later, a vast amount of research has been generated under the paradigm of cognitive linguistics, a flexible and evolving theoretical framework, bringing together several different projects. One of the main directions, in a way similar to the previous Chomskian paradigm, is Construction Grammar, where the basic unit of language is a conglomerate of syntactical, semantic and pragmatic information, instead of a schematic syntactic rule. Construction Grammar itself comprises several variations, from Cognitive Grammar (Langacker 1987), to Goldberg's Construction Grammar (Goldberg 2006), to Usage-Based Grammar (Tomasello 2005), just to mention a few.

Other directions explored aspects of language related to cognition that had been neglected in both the Chomskian tradition and in logic, such as lexical semantics. A well known difficulty in logical semantics is that of providing an account for the meaning of names, which include what has sometimes been called the extra-linguistic reference (Kripke 1972; Putnam 1975), or the referential component in lexical competence (Marconi 1997). Formal semantic models can only establish relations between symbols, that can build detailed inferences between concepts, but never attach components of meaning to the external world itself. Marconi (2000) has shown that even the most advanced model, that of Montague, fails to express a genuine reference to the world, and attaching labels to symbols does not solve the problem. From a cognitive perspective, it becomes natural to search for the referential component of lexical semantics, in the mental structures built on world experiences. How it works, and how experience relates to language, is a complex story (Violi 2001), but the road is well marked. Traveling quickly along that road we meet prototype theory of Rosch (1978), limited collections of exemplars representative of concepts; Lakoff (1987)'s radial categories around a central concept; Fillmore (1976)'s frames, conceptual systems where the understanding of any element requires the grasping of the structure as a whole. Specific phenomena within lexical semantics have also been analyzed by cognitive linguistics, such as polysemy (Tyler and Evans 2003), metaphor (Lakoff and Johnson 1980), and blending (Fauconnier 1997). Even if the cognitive linguistics approach to these phenomena has the merit of revealing important interactions between language and cognitive structures, it often remains shallow concerning the relations between both language and cognition and the external world.

One of the main tenets of cognitive linguistics is that it is a real "cognitive" theory, that attempts to describe language in connection to the rest of cognition, consistently with what other disciplines of cognitive science (e.g. neuropsychology, neural computation, developmental psychology) have revealed about cognition and the brain. A methodological example is the close connection with empirical studies on language acquisition (Elman et al. 1996; MacWhinney 1999; Bowerman and Levinson 2001; Diessel 2004).

Not much mathematics is found within cognitive linguistics, with just a few exceptions coming from Lakoff's school (Lakoff and Johnson 1999). Unfortunately, among the many fragments of semantic investigation, and of general intuitions concerning the concept of mental structures, nothing close to a common mathematical framework has emerged for a cognitively serious theory of meaning. It looks like the concerns of several logical semanticists on the lack of rigor of some of the ventures on the cognitive side of language were not ill posed, when freed from a consolidated and precise mathematical framework, any proposed subjective intuition can be given psychological reality without direct empirical verification.

This is one of the reasons for turning towards neurosemantics, as the only possible mathematical integration of linguistics with cognition, grounded on empirical bases. The distance to travel in order to reach a level of descriptive adequacy of logical semantics is huge indeed, but the target, in the long run, is much more ambitious: that of describing language not as an abstract external object, but rather as the way our brain represents the world.

# References


Baker, G., & Hacker, P. (1989). Frege's anti-psychologism. In M. A. Notturno (Ed.), *Perspectives on psychologism* (pp. 75–137). Leiden: Brill.

Biro, J. I., & Kotatko, P. (Eds.). (1995). *Frege, sense and reference one hundred years later*. Dordrecht: Kluwer.

Boole, G. (1854b), *An investigation of the laws of thought, on which are founded the mathematical theories of logic and probabilities*. London: Walton and Maberley (trad. it. di Mario Trinchero *Indagine sulle leggi del pensiero, cui sono fondate le teorie matematiche della logica e della probabilita'*, 1976).

Botha, R. P. (1999). On chomsky's "fable" of instantaneous language evolution. *Language & Communication, 19*, 243–257.

Bowerman, M., & Levinson, S. (Eds.). (2001). *Language acquisition and conceptual development*. Cambridge: Cambridge University Press.

Brdar, M., Gries, S., & Fuchs, M. (Eds.). (2011). *Cognitive linguistics – Convergence and expansion*. Amsterdam: John Benjamins.

Carnap, R. (1928). *Der logische Aufbau der Welt*. Berlin-Schlactensee: Weltkreis Verlag.

Chomsky, N. (1956). Three models for the description of languages. *IRE Transaction on Information Theory, 2*, 113–124.

Chomsky, N. (1957). *Syntactic structures*. The Hague: Mouton & Co.

Chomsky, N. (1958). On certain formal properties of grammars. *Information and Control, 1*, 91–112.

Chomsky, N. (1981). *Lectures in government and binding*. Dordrecht: Foris.

Chomsky, N. (1988). *Language and problems of knowledge: The managua lectures*. Cambridge: MIT.

Chomsky, N. (1993). A minimalist progam for linguistic theory. In K. Hale & S. J. Keyser (Eds.), *The view from building 20: Essays in linguistics in honor of Sylvain bromberger*. Cambridge: MIT.

Chomsky, N. (1995). *The minimalist program*. Cambridge: MIT.

Chomsky, N. (2000). *New horizons in the study of language and mind*. Cambridge: Cambridge University Press.

Chomsky, N. (2005). Three factors in language design. *Linguistic Inquiry, 36*, 1–22.

Chomsky, N., & Halle, M. (1968). *The sound pattern of English*. New York: Harper and Row

Chomsky, N., & Lasnik, H. (1993). The theory of principles and parameters. In J. Jacobs, A. von Stechow, W. Sternefeld, & T. Vennemann (Eds.), *Syntax – An international handbook of contemporary research*. Berlin: W. de Gruyter.

Croft, W., & Cruse, A. (2004). *Cognitive linguistics*. Cambridge: Cambridge University Press.

Diessel, H. (2004). *The acquisition of complex sentences*. Cambridge: Cambridge University Press.

Dummett, M. A. (1973). *Frege: Philosophy of language*. London: Duckworth.

Elman, J. L., Bates, E., Johnson, M. H., Karmiloff-Smith, A., Parisi, D., & Plunkett, K. (1996). *Rethinking innateness a connectionist perspective on development*. Cambridge: MIT.

Fauconnier, G. (1997). *Mappings in thought and language*. Cambridge: Cambridge University Press.

Fillmore, C. (1976). Frame semantics and the nature of language. *Annals of the New York Academy of Sciences, 280*, 20–32.

Fodor, J. (1983). *Modularity of mind: An essay on faculty psychology*. Cambridge: MIT.

Frege, G. (1879). *Begriffsschrift, eine der arithmetischen nachgebildete Formelsprache des reinen Denkens*. Halle: Louis Nebert.

Frege, G. (1884). *Die Grundlagen der Arithmetik: eine logisch-mathematische Untersuchung über den Begriff der Zahl*. Breslau: W. Koebner. (Reprinted by Olms, Hildesheim, 1961).

Frege, G. (1892). Über Sinn und Bedeutung. *Zeitschrift für Philosophie und philosophische Kritik, 100*, 25–50 (Reprinted in Angelelli, I. (Ed.) (1967). *Kleine Schriften*. Hildesheim: Olms).

Frege, G. (1894). Rezension von: E. Husserl, Philosophie der Arithmetik iI. *Zeitschrift für Philosophie und philosophische Kritik 103*, 313–332. (Reprinted in Angelelli, I. (Ed.) (1967). *Kleine Schriften*. Hildesheim: Olms).
Frege, G. (1904). Was ist eine Funktion? In S. Meyer (Ed.), *Festschrift für Ludwig Boltzmann gewidmet zum sechzigsten Geburtstage* (pp. 656–666). Leipzig: A. Barth. (Reprinted in Angelelli, I. (Ed.) (1967). *Kleine Schriften*. Hildesheim: Olms; Trad. it. di M. Carapezza e G. Rigamonti in Poggi, S. (Ed.) (2002). *Le leggi del pensiero tra logica, ontologia e psicologia.* Il dibattito austro-tedesco (1830–1930). Milano: Edizioni Unicopli).
Geeraerts, D. (Ed.). (2006). *Cognitive linguistics: Basic readings*. Berlin: Mouton de Gruyter.
Goldberg A (2006) *Constructions at work. The nature of generalization in language*. Oxford: Oxford University Press.
Goldsmith, J. (1985). A principled exception to the coordinate structure constraint. In *Papers from the Twenty-First Regional Meeting, Part I*. Chicago: Chicago Linguistic Society
Harris, R. A. (1993). *The linguistics wars*. Oxford: Oxford University Press.
Harris, Z. S. (1951). *Methods in structural linguistics*. Chicago: Chicago University Press.
Hobbes, T. (1651). Leviathan. London, trad. it. di G. Nicheli, La Nuova Italia, 1976.
Hodgkin, A. L., & Huxley, A. F. (1952). A quantitative description of ion currents and its applications to conduction and excitation in nerve membranes. *Journal of Physiology, 117*, 500–544.
Katz, J. (1981). *Language and other abstract objects*. Totowa: Rowman and Littlefield.
Katz, J., & Postal, P. (1991). Realism vs. conceptualism in linguistics. *Linguistics and Philosophy, 14*, 515–554.
Kleene, S. C. (1956). Representation of events in nerve nets and finite automata. *Automata Studies, 34*, 3–41.
Kripke, S. A. (1972). Naming and necessity. In D. Davidson & G. H. Harman (Eds.), *Semantics of natural language* (pp. 253–355). Dordrecht: Reidel Publishing.
Kripke, S. A. (1982). *Wittgenstein on rules and private language: An elementary exposition.* Cambridge: Harvard University Press.
Lakoff, G. (1986). A principled exception to the coordinate structure constraint. In *Proceedings of the Twenty-First Regional Meeting, Chicago Linguistic Society*. Chicago: Chicago Linguistic Society.
Lakoff, G. (1987). *Women, fire and dangerous things. What categories reveal about the mind.* Chicago: Chicago University Press.
Lakoff, G., & Johnson, M. (1980). *Metaphors we live by*. Chicago: Chicago University Press.
Lakoff, G., & Johnson, M. (1999). *Philosophy in the flesh. The embodied mind and its challenge to western thought*. New York: Basic Books.
Langacker, R. W. (1987). *Foundations of cognitive grammar*. Stanford: Stanford University Press.
Leibniz, G. W. (1684). Abhdlg. ohne Überschrift Vorarb. z. allg. Charakteristik. Berlin, in *Die philosophischen Schriften* a cura di C. I. Gerhardt, 1875–1863.
Lettvin, J., Maturana, H., McCulloch, W., & Pitts, W. (1959). What the frog's eye tells the frog's brain. *Proceedings of IRE, 47*, 1940–1951.
MacWhinney, B. (Ed.). (1999). *The emergence of language* (2nd ed.). Mahwah: Lawrence Erlbaum Associates.
Marconi, D. (1997). *Lexical competence*. Cambridge: MIT. (ediz. it. *Competenza Lessicale*, Laterza, 1999).
Marconi, D. (2000). What is montague semantics? In L. Albertazzi (Ed.), *Meaning and cognition: A multidisciplinary approach*. Amsterdam: John Benjamins.
Margolis, E., & Laurence, S. (2007). The ontology of concepts – abstract objects or or mental representations? *Nous, 41*, 561–593.
McCulloch, W., & Pitts, W. (1943). A logical calculus of the ideas immanent in nervous activity. *Bulletin of Mathematical Biophysics, 5*, 115–133.
Minsky, M. L. (1954). Neural nets and the brain-model problem. PhD thesis, Princeton University.
Osterhout, L., Kim, A., & Kuperberg, G. R. (2007). The neurobiology of sentence comprehension. In M. Spivey, K. McRae, & M. Joanisse (Eds.), *The Cambridge handbook of psycholinguistics* (pp. 365–389). Cambridge: Cambridge University Press.

Peacocke, C. (1992). Sense and justification. *Mind, 101*, 793–816.
Piccinini, G. (2004). The first computational theory of mind and brain: A close look at McCulloch and Pitts's 'Logical calculus of ideas immanent in nervous activity'. *Synthese, 141*, 175–215.
Pitts, W., & McCulloch, W. (1947). How we know universals: The perception of auditory and visual forms. *Bulletin of Mathematical Biophysics, 9*, 115–133.
Plebe, A. (2004). *Il linguaggio come calcolo – dalla logica di Boole alle reti neuronali*. Roma: SCIENTIA, Armando Editore.
Post, E. (1921). Introduction to a general theory of elementary propositions. *Journal of Mathematics, 43*, 163–185.
Post, E. (1947). Recursive unsolvability of a problem of Thue. *Journal of Symbolic Logic, 12*, 1–11.
Prat, C. S., Keller, T. A., & Just, M. A. (2007). Individual differences in sentence comprehension: A functional magnetic resonance imaging investigation of syntactic and lexical processing demands. *Journal of Cognitive Neuroscience, 19*, 1950–1963.
Proverbio, A. M., Crotti, N., Zani, A., & Adorni, R. (2009). The role of left and right hemispheres in the comprehension of idiomatic language: An electrical neuroimaging study. *BMC Neuroscience, 10*, 116.
Pulvermüller, F. (2010). Brain embodiment of syntax and grammar: Discrete combinatorial mechanisms spelt out in neuronal circuits. *Brain and Language, 112*, 167–179.
Putnam, H. (1975). The meaning of "meaning". In H. Putnam, *Mind, language and reality* (Vol. 2, pp. 215–271). Cambridge: MIT.
Rashevsky, N. (1938). *Mathematical biophysics: Physico-mathematical foundations of biology*. Chicago: Chicago University Press.
Rosch, E. (1978). Principles of categorization. In E. Rosch & B. Lloyd (Eds.), *Cognition and categorization*. Mahwah: Lawrence Erlbaum Associates.
Russell, B. (1902). Letter to Frege. In J. van Heijenoort (Ed.) (1967), *From Frege to Gödel: A source book in mathematical logic 1879–1931*. Cambridge: Harvard University Press.
Russell, B. (1905). On denoting. *Mind, 14*, 479–493.
Russell, B. (1918). *Mysticism and logic and other essays*. London: George Allen & Unwin.
Seuren, P. (2004a). *Chomsky's minimalism*. Oxford: Oxford University Press.
Smith, L. (1986). *Behaviorism and logical positivism: A reassessment of the alliance*. Stanford: Stanford University Press.
Stich, S. P., & Ravenscroft, I. (1994). What is folk psychology? *Cognition, 50*, 447–468.
Stowe, L. A., Haverkort, M., & Zwarts, F. (2004). Rethinking the neurological basis of language. *Lingua, 115*, 997–1042.
Thue, A. (1906). Über unendliche Zeichenreihen. *Norske Vid Selsk Skr I Mat Nat Kl, 7*, 1–22.
Thue, A. (1912). Über die gegnseitige Lage gleiche Teile gewisser Zeichenreihen. *Norske Vid Selsk Skr I Mat Nat Kl, 10*, 1–67.
Tomasello, M. (2005). *Constructing a language: A usage-based theory of language acquisition*. Cambridge: Harvard University Press.
Turing, A. (1936). On computable numbers, with an application to the Entscheidungsproblem. *Proceedings of the London Mathematical Society, 42*, 230–265.
Tyler, A., & Evans, V. (2003). *The semantics of English prepositions: Spatial scenes, cognition and the experiential basis of meaning*. Cambridge: Cambridge University Press.
van Benthem, J., & ter Meulen, A. (Eds.). (1996). *Handbook of logic and language*. Amsterdam: Elsevier.
Vassallo, N. (2000). Psychologism in logic: Some similarities between boole and frege. In J. Gasser (Ed.), *A Boole anthology*. Berlin: Springer.
Violi, P. (2001). *Meaning and experience*. Bloomington: Indiana University Press.
von Neumann, J. (1958). *The computer and the brain*. New Haven: Yale University Press.
Wittgenstein, L. (1922b). *Tractatus Logico-Philosophicus*. London: Trench, Trubner & Co.
Wittgenstein, L. (1953). *Philosophische Untersuchung*. Oxford: Basil Blackwell.

# Chapter 6
# Neurosemantics of Visual Objects

**Abstract** Humans, like several other primates, are visual creatures, and almost half of our neurons are devoted to the processing of visual signals. The excellence found in our ability to do so, is not just due to our ophthalmological capabilities, which are outperformed by other species, such as birds, it is instead on the semantic side, in our ability to classify hundreds of object categories on the basis of their visual appearance only. Vision has historically been the earliest and most investigated function in the brain, thanks to its unique correspondence between the two dimensional organization of the distal stimulus and cortical processing units. Taken together, these two factors have led us to investigating the semantics of objects whose essential features are captured by their visual appearance. The first model presented in this chapter is a sort of prelude to a full blown semantics, with a simulation of the full visual pathway that brings light signals into recognition of object categories, together with the auditory pathway, in a simulation of the emergence of a first lexicon, that in infants begins exactly with visual objects. Most of the components of this model, and the methods used for its development and subsequent analyses, will be shared by the models that follow. The second model presented in this chapter, taps into a range of semantic phenomena typically observed in the early stages of language development in children, such as the change in the speed of learning, and the so called "fast-mapping" phenomenon.

## 6.1 Object Recognition

The first neurosemantic model presented in this section is based on visual objects. Mammals rely a great deal on their visual system and as a result, one of the primary sources of information on the external world arrives in the form of visual input. One of the challenges our visual perception presents in building a representation of the world in our brain, is that of learning which light sensations belong to the same class of entities, despite significant changes in appearance. In humans, this effort is soon intensified when yet another precious but different source of sensory information is added to the mix, that is, when the mind grasps the idea that often, patterns of sound are used to identify and categorize visible objects. This event takes place in the brain at a boundary between the visual system and the language system, and it is the target of this first model.

 
A. Plebe, V.M. De La Cruz, *Neurosemantics*,
Studies in Brain and Mind 10, DOI 10.1007/978-3-319-28552-8_6

It is not by chance that the initial vocabulary acquired by very young children is made up to a large extent by nouns referring to visible objects (Bates et al. 1995; Gershkoff-Stowe and Smith 2004). In a cross-linguistic study from seven different linguistic communities (Argentina, Belgium, France, Israel, Italy, the Republic of Korea, and the United States), Bornstein and R.Cote (2004) found that children with vocabularies of 51–100 and 101–200 words had more nouns than any other grammatical class. The formation of the ability to segment and recognize objects on one side, and to segment and recognize words on the other, and the ability to join the two represented entities in linguistic meaning, takes place in the brain at the crossroad between the ventral visual and auditory streams of processing. This is the brain portion simulated in this model.

There have been assorted attempts to investigate how the human mind acquires the mapping between words and categories of objects, by means of computational models. Some of them belong to the connectionist computational paradigm. Rogers and McClelland (2006) explored the building of simple conceptual systems, using the standard PDP framework (Rumelhart and McClelland 1986b). Their model learns by backpropagation, categories such as `flower`, `tree`, in correlation with visual features, such as `red` or `branches`, together with a fixed set of attributes, like `can walk`, `is living`. Despite the higher level of abstraction, and the lack of visual features proper, this model simulates important phenomena, for example the emergence of quite general dimensions of similarity without appealing to either physiological or cognitive constraints, but simply as the result of a coherent covariation of features. The conceptual system of Rogers and McClelland by itself does not imply a lexical semantics, which instead is the target of the LEX model by Regier (2005). The semantic content in this model, is deprived from any conceptual constituents, most of the focus of the model is on the association with the phonological form of the words. This too is a connectionist model and the level of abstraction is high, with both phonological and semantic features predefined in a conventional way, without any relation to real utterances. A similar approach is pursed by Mayor and Plunkett (2010), since their model explores the same specific aspect of lexical categorization that is the focus of our next model, it will be discussed in the section pertaining to it (see Sect. 6.2).

Other computational models exist that also aim at understanding visual object recognition, with some being inspired by realistic brain processes (Edelman and Duvdevani-Bar 1997; Riesenhuber and Poggio 2000). There are a couple of examples where, to some extent, the hierarchy of the visual cortex had been reproduced (Wallis and Rolls 1997; Deco and Rolls 2004; Rolls and Stringer 2006; Taylor et al. 2005). Very few models extend beyond the occipital cortex, but Kashimori et al. (2007) is an example of one that proposed a neural model that includes ITC (Inferior Temporal Cortex) and PFC (PreFrontal Cortex), giving an account of their different roles in categorization. The ITC response is much more influenced by visual features than PFC, even if only by those features important for categorization, and the response in PFC is sustained even after the disappearance of the visual stimulus. Note that the hierarchy of maps of this model reflects the organization of the visual system in monkeys, which differs from that of humans especially in the higher areas, as discussed in Sect. 3.4.1. A recent comparison of several models of purely visual categorization is in (Khaligh-Razavi and Kriegeskorte 2014). The model here described derives from previous developments on visual object recognition (Plebe and Domenella 2005, 2006, 2007).

Not many neural models have been proposed for auditory processes (Näger et al. 2002; Volkmer 2004), and little is yet known about the kind of brain computations that lead to word recognition there. The linguistic integration inside this model has been developed in several stages (Plebe 2007b; Plebe et al. 2007, 2010, 2011).

In Sect. 3.2 we introduced the mechanism of coincidence detection as, in our opinion, the fundamental principle of cortical representation. We take it as the core mechanism in this model for explaining the emergence of the semantics of visual objects. Coincidence detection, as argued in Sect. 3.2, is a multilevel mechanism. At the lower level, it is implemented in local synaptic connections by their sensitivity to the occurrence of simultaneous activation of neighboring units. At an intermediate level, it is responsible for building selectivity in units to recurrent patterns, such as oriented lines in the visual scene, or classes of phonemes. At the highest level, it captures the coincidence of seeing certain objects while hearing the same sound, which then becomes associated with the category of similar objects named by the sound, which is a word. We will certainly concede that in reality, the formation of object semantics in humans is a complex event, where many different mechanisms converge. For example, Tomasello (1999) pinpointed the role of pragmatic and social cues in grasping the association between uttered words and objects, something that has been confirmed in several studies (Grassman et al. 2009).

An emerging body of new evidence coming from recent developmental studies takes into consideration the child's very own visual perspective, using an embodied approach. These investigations use head cameras mounted on toddlers' heads, and their results indicate that word learning in 18 month olds occured more efficiently when bottom-up visual information was "clean and uncluttered" (Yu and Smith 2012; Pereira et al. 2014). In other words, when objects where in optimal view, from the child's perspective, during the naming event. The optimal viewing conditions were determined by the child's own interaction with objects and with her parents, during the naming event. This work has important implications in that while many theories on word learning invoke the referential ambiguity that is an integral part of learning a word when it co-occurs with a natural scene, it shows how one of the mechanisms that might be helping to resolve this ambiguity is the child's own sensory-motor behavior and that of her parents in producing "optimal visual moments", facilitating not only early object recognition but early word learning.

Our model is not in contrast with these positions, it simply deals with a part of the overall sequence involved in a naming experience. We can assume that social cues drive attentional mechanisms, that filter, among all the possible objects presented in a scene, the one focused on by the speaker. We can further assume that, among this filtered set of experiences, additional filtering provides a smaller set of samples by using "optimal visual moments". At this point the framed object becomes the input of our model, which is restricted to the primary aspect of learning that a sound conveys meaning related to a category of similar objects.

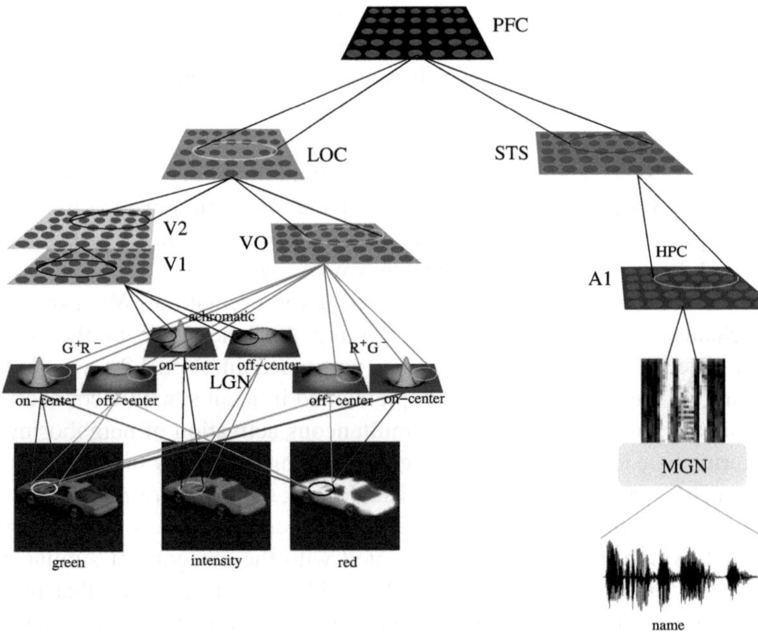


**Fig. 6.1** Overall scheme of the object categorization model, the acronyms of all components are in Table 6.1

**Table 6.1** Main parameters of the object categorization model

| Map | | Dimension | $r_A$ | $r_E$ | $r_I$ | $\gamma_A$ | $\gamma_E$ | $\gamma_I$ | $\gamma_N$ |
|---|---|---|---|---|---|---|---|---|---|
| LGN | Lateral geniculated nucleus | 144 × 144 | 4.7 | – | – | – | – | – | – |
| MGN | Medial geniculated nucleus | 32 × 32 | 2.0 | – | – | – | – | – | – |
| V1 | Primary visual cortex | 96 × 96 | 8.5 | 1.5 | 7.0 | 1.5 | 1.0 | 1.0 | 0.0 |
| V2 | Secondary visual cortex | 30 × 30 | 7.5 | 8.5 | 3.5 | 50.0 | 3.2 | 2.5 | 0.7 |
| VO | Ventral occipital | 30 × 30 | 24.5 | 4.0 | 8.0 | 1.8 | 1.0 | 1.0 | 0.0 |
| A1 | Auditory primary cortex | 24 × 24 | 6.5 | 1.5 | 5.0 | 1.7 | 0.9 | 0.9 | 0.0 |
| LOC | Lateral occipital complex | 16 × 16 | 6.5 | 1.5 | 3.5 | 0.5 | 1.1 | 1.7 | 0.0 |
| STS | Superior temporal sulcus | 16 × 16 | 2.5 | 1.5 | 5.5 | 1.8 | 1.0 | 1.2 | 0.0 |
| PFC | Pre-frontal cortex | 24 × 24 | 2.5 | 1.5 | 5.5 | 1.5 | 3.2 | 4.1 | 0.0 |

### 6.1.1 *The Cortical Maps Structure*

The model is built upon a number of Topographica maps, simulating several cortical areas, as well as on a few thalamic maps, shown in Fig. 6.1, the legend of all maps, together with their main parameters, is provided in Table 6.1. The structure of the model is based on the accepted perspective that both the visual and the auditory processing pathways in the brain can be divided into two broad main streams, as extensively discussed in Sect. 3.4, therefore, for the purpose of exploring semantics, it suffices to include only the ventral areas.

Both visual and auditory paths in the model include thalamic nuclei, which perform their own processing of the signals coming from sensory receptors. Given that their maturation is already advanced at the age relevant for the emergence of semantics, and that their detailed functions are not relevant in the scope of this study, all subcortical processes are hardwired, according to the following equations:

$$x^{\odot} = f\left(\left(\mathbf{l}_{r_A} + \mathbf{m}_{r_A}\right) \cdot \left(\mathbf{g}_{r_A}^{\sigma_N} - \mathbf{g}_{r_A}^{\sigma_W}\right)\right) \tag{6.1}$$

$$x^{\circledcirc} = f\left(\left(\mathbf{l}_{r_A} + \mathbf{m}_{r_A}\right) \cdot \left(\mathbf{g}_{r_A}^{\sigma_W} - \mathbf{g}_{r_A}^{\sigma_N}\right)\right) \tag{6.2}$$

$$x^{R^+G^-\odot} = f\left(\mathbf{l}_{r_A} \cdot \mathbf{g}_{r_A}^{\sigma_N} - \mathbf{m}_{r_A}\mathbf{g}_{r_A}^{\sigma_W}\right) \tag{6.3}$$

$$x^{R^+G^-\circledcirc} = f\left(\mathbf{l}_{r_A} \cdot \mathbf{g}_{r_A}^{\sigma_W} - \mathbf{m}_{r_A}\mathbf{g}_{r_A}^{\sigma_N}\right) \tag{6.4}$$

$$x^{G^+R^-\odot} = f\left(\mathbf{m}_{r_A} \cdot \mathbf{g}_{r_A}^{\sigma_N} - \mathbf{l}_{r_A}\mathbf{g}_{r_A}^{\sigma_W}\right) \tag{6.5}$$

$$x^{G^+R^-\circledcirc} = f\left(\mathbf{m}_{r_A} \cdot \mathbf{g}_{r_A}^{\sigma_W} - \mathbf{l}_{r_A}\mathbf{g}_{r_A}^{\sigma_N}\right) \tag{6.6}$$

$$x_{\tau,\omega}^{MGN} = \left| \sum_{t=t_0}^{t_M} v(t) w(t-\tau) e^{-j\omega t} \right|^2 \tag{6.7}$$

where $x$ is the activity of a unit. The subscript has been omitted for clarity in all equations except for (6.7), where $\tau$ is the time dimension, and $\omega$ is the sound frequency dimension. In Eq. (6.7) $w(\cdot)$ is a small time window, that performs a spectrogram-like response, of the type accomplished by the combination of cochlear and MGN nucleus processes (Brown 2003). In the visual path there are separated LGN sheets for the achromatic component, described by Eqs. (6.1) and (6.2), and other sheets for the chromatic components, for the medium and long wavelength, with equations (6.3), (6.4), (6.5), (6.6). The superscript $\odot$ marks receptive fields with the inner part active, while the superscript $\circledcirc$ makes those with a peripheral ring active .

The profile of the visual receptive fields is given by differences of two Gaussian $\mathbf{g}^{\sigma_N}$ and $\mathbf{g}^{\sigma_W}$, with standard deviation $\sigma_N < \sigma_W$, approximating the combined contribution of gangliar cells and LGN (Dowling 1987). The chromatic receptive fields combine the center/periphery response with the opponency of two colors, as shown in Fig. 6.1. For example the response of the unit $x^{R^+G^-\odot}$ will be maximized by a central red spot surrounded by green.

The maps corresponding to early cortical processes are V1, V2, VO, and A1, here follow their equations:

$$x^{V1} = f\Big(\gamma_A^{V1}\left(\mathbf{a}_{r_A}^{V1\leftarrow\odot} \cdot \mathbf{x}_{r_A}^{\odot} + \mathbf{a}_{r_A}^{V1\leftarrow\circledcirc} \cdot \mathbf{x}_{r_A}^{\circledcirc}\right) + \gamma_E^{V1}\mathbf{e}_{r_E}^{V1} \cdot \mathbf{x}_{r_E}^{V1} - \gamma_I^{V1}\mathbf{i}_{r_I}^{V1} \cdot \mathbf{x}_{r_I}^{V1}\Big) \tag{6.8}$$

$$x^{\mathrm{V2}} = f\left(\gamma_{\mathrm{A}}^{\mathrm{V2}} \mathbf{a}_{r_{\mathrm{A}}}^{\mathrm{V2}\leftarrow V1} \cdot \mathbf{x}_{r_{\mathrm{A}}}^{V1} + \gamma_{\mathrm{E}}^{\mathrm{V2}} \mathbf{e}_{r_{\mathrm{E}}}^{\mathrm{V2}} \cdot \mathbf{x}_{r_{\mathrm{E}}}^{\mathrm{V2}} - \gamma_{\mathrm{I}}^{\mathrm{V2}} \mathbf{i}_{r_{\mathrm{I}}}^{\mathrm{V2}} \cdot \mathbf{x}_{r_{\mathrm{I}}}^{\mathrm{V2}}\right) \tag{6.9}$$

$$\begin{aligned} x^{\mathrm{VO}} = f\Bigg( & \gamma_{\mathrm{A}}^{\mathrm{VO}} \Big( \mathbf{a}_{r_{\mathrm{A}}}^{\mathrm{VO}\leftarrow \mathrm{R^+G^-}\odot} \cdot \mathbf{x}_{r_{\mathrm{A}}}^{\mathrm{R^+G^-}\odot} + \mathbf{a}_{r_{\mathrm{A}}}^{\mathrm{VO}\leftarrow \mathrm{R^+G^-}\circledcirc} \cdot \mathbf{x}_{r_{\mathrm{A}}}^{\mathrm{R^+G^-}\circledcirc} \\ & + \mathbf{a}_{r_{\mathrm{A}}}^{\mathrm{VO}\leftarrow \mathrm{G^+R^-}\odot} \cdot \mathbf{x}_{r_{\mathrm{A}}}^{\mathrm{G^+R^-}\odot} + \mathbf{a}_{r_{\mathrm{A}}}^{\mathrm{VO}\leftarrow \mathrm{G^+R^-}\circledcirc} \cdot \mathbf{x}_{r_{\mathrm{A}}}^{\mathrm{G^+R^-}\circledcirc} \Big) \\ & + \gamma_{\mathrm{E}}^{\mathrm{VO}} \mathbf{e}_{r_{\mathrm{E}}}^{\mathrm{VO}} \cdot \mathbf{x}_{r_{\mathrm{E}}}^{\mathrm{VO}} - \gamma_{\mathrm{I}}^{\mathrm{VO}} \mathbf{i}_{r_{\mathrm{I}}}^{\mathrm{VO}} \cdot \mathbf{x}_{r_{\mathrm{I}}}^{\mathrm{VO}} \Bigg) \end{aligned} \tag{6.10}$$

$$x^{\mathrm{A1}} = f\left(\gamma_{\mathrm{A}}^{\mathrm{A1}} \mathbf{a}_{r_{\mathrm{A}}}^{\mathrm{A1}\leftarrow MGN} \cdot \mathbf{x}_{r_{\mathrm{A}}}^{MGN} + \gamma_{\mathrm{E}}^{\mathrm{A1}} \mathbf{e}_{r_{\mathrm{E}}}^{\mathrm{A1}} \cdot \mathbf{x}_{r_{\mathrm{E}}}^{\mathrm{A1}} - \gamma_{\mathrm{I}}^{\mathrm{A1}} \mathbf{i}_{r_{\mathrm{I}}}^{\mathrm{A1}} \cdot \mathbf{x}_{r_{\mathrm{I}}}^{\mathrm{A1}}\right) \tag{6.11}$$

It can be easily seen that all equations are direct derivations from (4.13), adapted according to the placement of every map in the overall hierarchy. In the visual part we meet the same areas described Sect. 3.4.1, here area V3 is not included, since the experiment deals with static scenes only. A technical simplification is to segregate the form processing in V1 and V2, using Eqs. (6.8) and (6.9), and the color processing in VO, with Eq. (6.10).

The higher level of cortical process is simulated by the following equations:

$$\begin{aligned} x^{\mathrm{LOC}} = f\Bigg( & \gamma_{\mathrm{A}}^{\mathrm{LOC}} \left( \mathbf{a}_{r_{\mathrm{A}}}^{\mathrm{LOC}\leftarrow V2} \cdot \mathbf{x}_{r_{\mathrm{A}}}^{V2} + \mathbf{a}_{r_{\mathrm{A}}}^{\mathrm{LOC}\leftarrow VO} \cdot \mathbf{x}_{r_{\mathrm{A}}}^{VO} \right) \\ & + \gamma_{\mathrm{E}}^{\mathrm{LOC}} \mathbf{e}_{r_{\mathrm{E}}}^{\mathrm{LOC}} \cdot \mathbf{x}_{r_{\mathrm{E}}}^{\mathrm{LOC}} - \gamma_{\mathrm{I}}^{\mathrm{LOC}} \mathbf{i}_{r_{\mathrm{I}}}^{\mathrm{LOC}} \cdot \mathbf{x}_{r_{\mathrm{I}}}^{\mathrm{LOC}} \Bigg) \end{aligned} \tag{6.12}$$

$$\begin{aligned} x^{\mathrm{STS}} = f\Bigg( & \gamma_{\mathrm{A}}^{\mathrm{STS}} \mathbf{a}_{r_{\mathrm{A}}}^{\mathrm{STS}\leftarrow A1} \cdot \mathbf{x}_{r_{\mathrm{A}}}^{A1} + \gamma_{\mathrm{E}}^{\mathrm{STS}} \mathbf{e}_{r_{\mathrm{E}}}^{\mathrm{STS}} \cdot \mathbf{x}_{r_{\mathrm{E}}}^{\mathrm{STS}} \\ & - \gamma_{\mathrm{I}}^{\mathrm{STS}} \mathbf{i}_{r_{\mathrm{I}}}^{\mathrm{STS}} \cdot \mathbf{x}_{r_{\mathrm{I}}}^{\mathrm{STS}} \Bigg) \end{aligned} \tag{6.13}$$

$$\begin{aligned} x^{\mathrm{PFC}} = f\Bigg( & \gamma_{\mathrm{A}}^{\mathrm{PFC}} \left( \mathbf{a}_{r_{\mathrm{A}}}^{\mathrm{PFC}\leftarrow \mathrm{LOC}} \cdot \mathbf{x}_{r_{\mathrm{A}}}^{\mathrm{LOC}} + \mathbf{a}_{r_{\mathrm{A}}}^{\mathrm{PFC}\leftarrow \mathrm{STS}} \cdot \mathbf{x}_{r_{\mathrm{A}}}^{\mathrm{STS}} \right) \\ & + \gamma_{\mathrm{E}}^{\mathrm{PFC}} \mathbf{e}_{r_{\mathrm{E}}}^{\mathrm{PFC}} \cdot \mathbf{x}_{r_{\mathrm{E}}}^{\mathrm{PFC}} - \gamma_{\mathrm{I}}^{\mathrm{PFC}} \mathbf{i}_{r_{\mathrm{I}}}^{\mathrm{PFC}} \cdot \mathbf{x}_{r_{\mathrm{I}}}^{\mathrm{PFC}} \Bigg) \end{aligned} \tag{6.14}$$

The LOC Map, as seen in Sect. 3.4.1, plays a crucial role for object recognition in humans, and it is ruled by Eq. (6.12), where chromatic and early form processing converge. Equation (6.13) governs the projection from A1 to STS, where basic phonological recognition takes place, as seen in Sect. 3.4.2. The highest map in the model, where auditory and visual information meet, is PFC, and follows Eq. (6.14). Certainly the more dramatic simplification of the model is to bind the final semantic processing inside a single map. It is well known that the semantic coding of visual objects is spread throughout the brain. However, there is also ample evidence that sustains that the PFC area is deeply engaged in the kind of semantic representation here investigated (Freedman et al. 2001, 2003; Miller et al. 2002; Wood and Grafman 2003; Ashby and Spiering 2004; Huey et al. 2006; Fuster 2008; Fairhall and Caramazza 2013).

### 6.1.2 Simulation of Experiences

Initially, the model lacks functional processing, with all synaptic connections initialized to small random values. During the experiments the model is exposed to a series of stimuli, which reproduce at a small and essential scale, early human development relevant for its cortical areas.

In a first phase only maps V1, VO, and A1 are plastic, and adapt their synaptic connections according to Eqs. (4.10), (4.11), and (4.12). The visual stimuli are synthetic random blobs that mimic waves of spontaneous retinal activity, that are known to play a fundamental role in the ontogenesis of the visual system (Mastronarde 1983; Katz and Shatz 1996; Thompson 1997; Gödecke and Bonhoeffer 1996; Chapman et al. 1996). Those presented to V1 are elongated along random directions, to stimulate orientation selectivity, while blobs to VO are circular, with constant hues, and random size, position, and intensity, in order to induce color constancy. The A1 map is exposed to short trains of waves sweeping linearly around a central frequency, with time duration, central frequencies and sweeping intervals varied randomly.

The second phase involves V2 and STS maps as well. The visual stimuli comprises pairs of elongated blobs, the same previously used for V1, with a coinciding end point. These sort of patterns stimulate the selectivity of units to patterns that are slightly more complex than oriented lines, like corners. The auditory stimuli are synthesized waves of the 7200 most common English words with length in range from 3 to 10 characters, generated using *Festival* (Black and Taylor 1997), tuned at cepstral order 64 and 2.3 s time window.

In a third phase, that corresponds to the phase just after eye opening in the newborn, natural images are used. In order to include the identification of an object under different perspectives, the COIL-100 collection has been used (Nayar and Murase 1995), where for each of the 100 real childhood related objects, 72 different views are available.

The last phase corresponds to early language acquisition, and the model is exposed to events in which an object is viewed and a label corresponding to its basic category is heard simultaneously. The 100 objects have been grouped manually into 38 categories. Certain categories, such as `cup` or `medicine` have five exemplars in the object collection, while others, such as `telephone`, have only one exemplar. Each category word is converted from text to waves using the `en1` "Roger" male voice, and the `us1` female American speaker in the *Festival* software. Both male and female utterances are duplicated at standard and slower speeds, using the 1.3 value of the `Duration_Stretch` parameter in *Festival*. Examples of all stimuli used in this model are in Fig. 6.2.

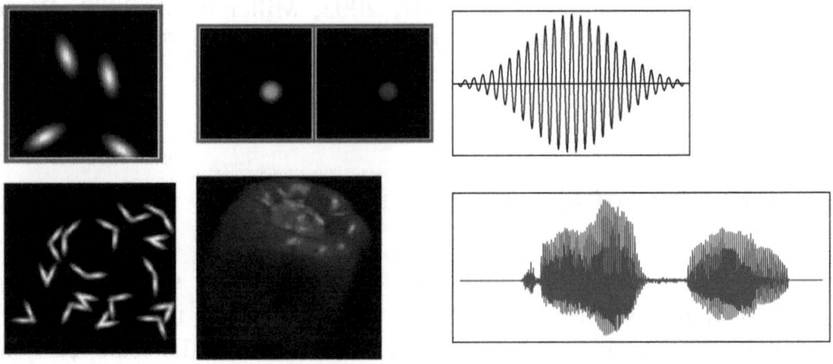

**Fig. 6.2** Samples of the stimuli presented to the model. In the *top row*, from the left, elongated blobs input to V1, hue circular blobs for VO, wave trains for A1. In the *bottom row*, from the left, couple of blobs for V2, real images for the visual path, and word waves for the auditory path

We would like to comment on the realism of the experiences used in developing this model, the same considerations hold for most of the further neurosemantic models that will be presented in this book. All models are composed by a complex, yet extremely simplified with respect to the real brain, hierarchy of neural sheets. In each, the emergence of functions consistent with their physiological counterpart areas is expected, and that it will contribute to the subsequent sheets, up to the higher level. For this purpose, often the range of stimuli used at the earlier stages of development are a subset of all the possible stimuli that may achieve similar functions, in order to simplify the overall experiment. It is clearly beyond the scope of studies focused on neurosemantics, to analyze in detail the dependency of each lower level component on the stimuli regimen. However, other studies using the same Topographica architecture have already addressed these details in depth. For example, the prenatal experience for V1 in this model includes random elongated blobs, which are the optimal type of patterns for stimulating the emergence of orientation selectivity in V1. Previous studies demonstrated that similar results can be achieved with *any* set of patterns, provided they are rich enough in variety of edges: synthetic disks, man-made objects, natural objects (flower and plants), landscapes, and faces (Bednar and Miikkulainen 2004; Plebe and Domenella 2007). No orientation develops when using random noise ad high frequency. Skewed data sets were indeed reflected in the maps, for example with landscapes the orientation histogram became biased towards horizontal, while using faces more neurons became tuned to vertical orientations, replicating findings from animals raised in biased environments. Similarly, in this model experiences for V2 are combinations of two random oriented elongated blobs, optimal for the development of angle selectivity. The same responsiveness has been studied using patterns that include single synthetic blobs, selections of sharp-edged objects, selection of rounded objects, white noise (Plebe 2007a). Only white noise prevented the development of angle selectivity.

### 6.1.3 Lexical Categorization

At the end of the simulated development, several types of topological organization can be found in the maps of the model, which are consistent with the known role of those maps in cortical hierarchy. The V1 map is organized basically with respect to orientation selectivity, in the VO map most units respond to specific hues, regardless of intensity, and the V2 map becomes responsive mainly to angles. Further details on the functions that emerge in the lower areas, not included here, can be found in (Plebe and Domenella 2007) concerning V1 and VO, and in (Plebe 2007a, 2012) for V2.

The analysis of the higher maps is carried out under the assumption of population coding, introduced in Sect. 4.3, using the algorithm described in Sect. 4.3. By deriving Eq. (4.15) for the LOC map, and using as stimulus the view $o$ of object $O$, we obtain the following:

$$x_i^{\mathrm{LOC}}(o) : O \in \mathscr{O} \rightarrow \mathbb{R}^{+}; \quad o \in O \in \mathscr{O}, \tag{6.15}$$

where $x_i^{\mathrm{LOC}}(o)$ is computed by (6.12), when the image $o$ is presented as input to the visual system. In this case, a category $S$ of (4.15) is just an individual object, whose instances are its possible views, which can be understood as belonging to the same entity, or mistaken for different objects, in which case a peculiar view would mislead. It is a type of perceptual error that happens in humans (Farah 1990). Examples of population coding in LOC are given in Fig. 6.3.

We can see the degree of invariance in the LOC responses to different views of the same object, with most units responding independently from the specific view of the individual object. As described in Sect. 3.4.1, the human LOC does not exhibit absolute invariance to possible transformations of the same object, rather a degree of tolerance with respect to classes of changes in the appearance of the same object. By using Eqs. (4.20), adapted to stimuli conditions described in (6.15), the amount of discrimination performed by the population coding in LOC has been quantitatively assessed, and is shown in Table 6.2.

The STS map codes words for their phonological form, now the basic Eq. (4.15) becomes the following:

$$x_i^{\mathrm{STS}}(n) : N \in \mathscr{N} \rightarrow \mathbb{R}; \quad n \in N \in \mathscr{N}, \tag{6.16}$$

where $x_i^{\mathrm{STS}}(n)$ is computed by (6.13), when the sound $n$ is presented to the auditory path. There are 38 classes $N$ of sounds. Examples of population coding for some of the words used as object labels are given in Fig. 6.4, Table 6.3 shows the accuracy of STS in discriminating between all the different words known by the model. In developing the functions in the STS map three different sets of stimuli have been experimented: female voices only, male voices only, or using the full set of voices. Listening to voices of mixed genders makes the identification of names more difficult, a phenomenon that has been observed in children.

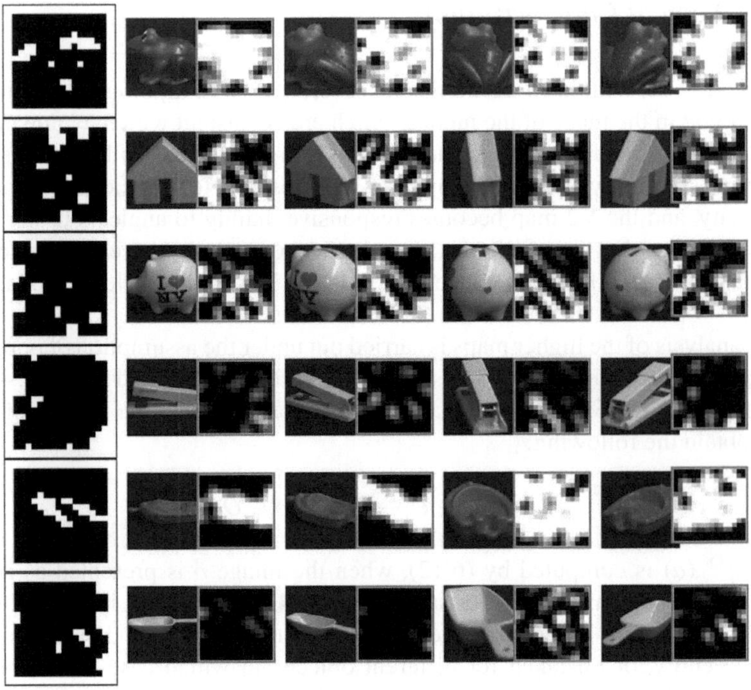

**Fig. 6.3** Examples of population coding in the LOC map. In each row the map on the far left displays the coding units. The following images in the row, are samples of views of the same object, and the corresponding response patterns in LOC

**Table 6.2** Ability of LOC map in discriminating individual objects by population coding. As a comparative figure, discrimination by chance would be 0.01

| | Discrimination ability | |
|---|---|---|
| # of view | Mean over objects | Standard deviation |
| 4 | 0.624 | 0.325 |
| 8 | 0.647 | 0.314 |
| 18 | 0.653 | 0.323 |

The derivation from the basic Eq. (4.15) for the PFC map is the following:

$$x_i^{\mathrm{PFC}}(c) : C \in \mathscr{C} \to \mathbb{R}; \quad c = \langle o, n \rangle \in C = \left( \{\epsilon\} \cup \bigcup_{O \in \mathscr{O}_C} O \right) \times (\{\epsilon\} \cup N_C), \tag{6.17}$$

where $x_i^{\mathrm{PFC}}(c)$ is computed by (6.14), when the sound $n$ is presented to the auditory path, and in coincidence the object $o$ is presented to the visual path. The 38 object categories introduce a partition in the set of objects $\mathscr{O}$, such that all sets of views in the partition $O \in \mathscr{O}_C$ are of objects of that category $C$. $N_C$ is the set of utterances

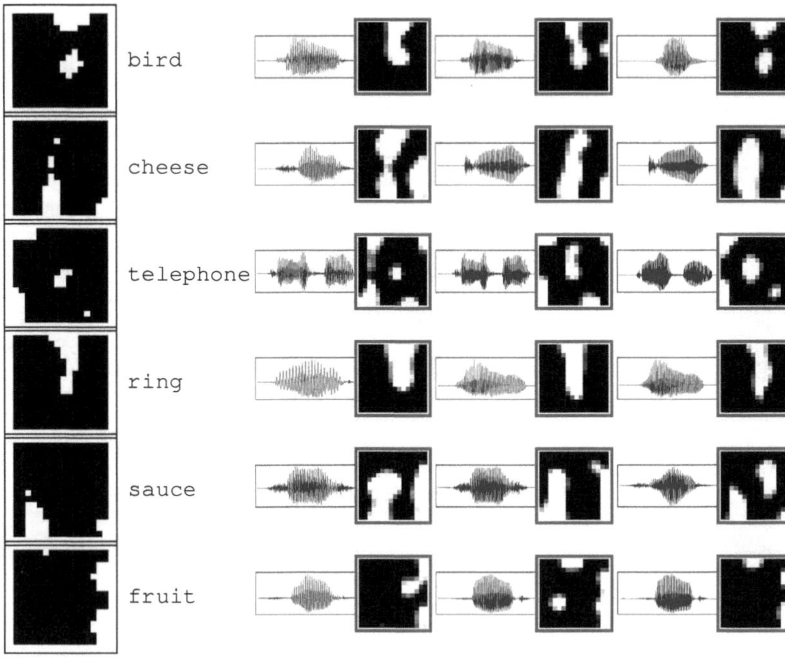


**Fig. 6.4** Examples of population coding of word sounds in the STS map. In each row the leftmost map displays the coding units. The other images in the row are samples of sound of the same label, with corresponding response patterns in STS

**Table 6.3** Discrimination ability of words, heard by different voices, in map STS. As a comparative figure, discrimination by chance would be 0.026

| | Discrimination ability | |
|---|---|---|
| Voices | Mean over names | Standard deviation |
| Female | 0.882 | 0.242 |
| Male | 0.895 | 0.234 |
| Both | 0.658 | 0.300 |

**Table 6.4** Accuracy of the map PFC in lexical categorization by population coding. As a comparative figure, discrimination by chance would be 0.026

| | Discrimination ability | |
|---|---|---|
| Voices | Mean over names | Standard deviation |
| Female | 0.878 | 0.220 |
| Male | 0.895 | 0.167 |
| Both | 0.695 | 0.240 |

naming category $C$. Note that the empty sample $\epsilon$ is included, for experiments in which only a single modality is presented. Therefore, $c = \langle o, \epsilon \rangle$ is the case of the visual modality only, and $c = \langle \epsilon, n \rangle$ is the case of linguistic input only.

In Table 6.4 there are the accuracy values obtained in map PFC, in recognizing objects of a given category. As for STS, the discrimination is quite more accurate when words are spoken by a person of a single gender, nevertheless, the model

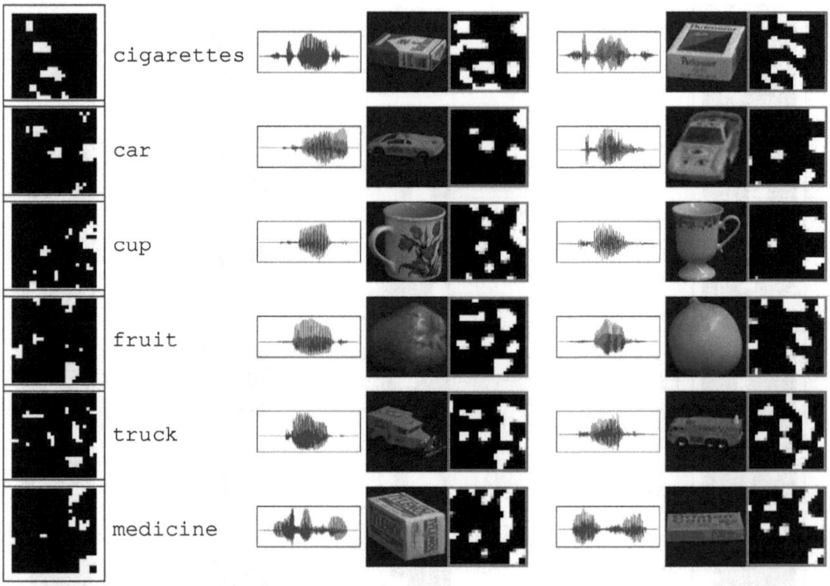


**Fig. 6.5** Examples of population coding in map PFC. In each row the leftmost map displays the coding units. Next, are the images that are samples of a sound of the same name, a view of an object of the relevant class, and the corresponding response patterns in PFC. The first sample always uses a male voice, and the second sample a female voice

achieves a remarkable ability of categorizing objects taking into account their names in all possible conditions. It is interesting to note, as shown in these images, that the coding does not reflect any explicit trace of visual features, which were still present in the LOC map. At this level retinotopy disappears, and the topological ordering is at an abstract level. Moreover, we can see how there are important overlaps between the category coding (the leftmost map) and contingent activation, with just small differences. The few units that depart from the category coding are those responding to peculiarities of a specific view of the object, or of the specific voice heard. Codings shown in Fig. 6.5 are amodal in kind, and associate information of a visual and acoustic nature. It is possible to analyze the two contributions separately, testing the model with two sorts of stimuli: $c = \langle o, \epsilon \rangle$ or $c = \langle \epsilon, n \rangle$, and always using Eqs. (4.18) and (4.20).

Some examples resulting from this analysis are shown in Fig. 6.6, where the separated coding of visual and auditory information is compared with the full amodal population coding. It is interesting to detect a global partitioning in map PFC, with linguistic information more clustered on the right side and visual on the left, still with large overlaps. Often the amodal coding looks like a combination of the visual and linguistic representations, as in the case of `car`, `fruit`, `pepper`. In other cases, one modality seems to be prevalent, for example `telephone` is more influenced by the linguistic representation, while for `plug` the visual information

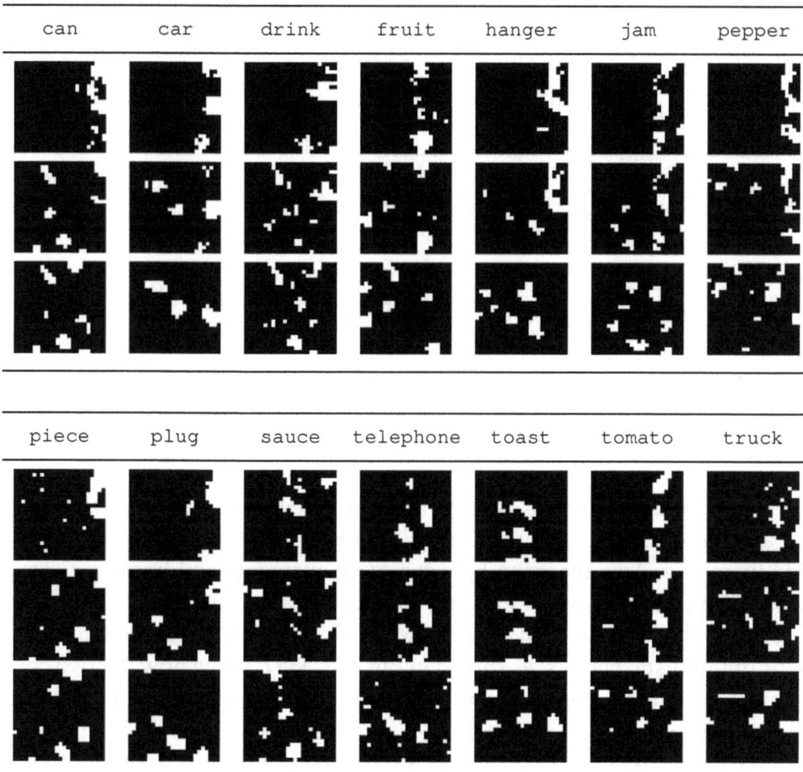


**Fig. 6.6** Examples in the PFC map of separated coding of visual and auditory information. Images in the *upper row* are coding maps for the auditory mode only, in the *lower row* we find the visual mode only, and the *middle row* is the full amodal population coding of the categories

is prevailing. This could be due, in the case of `telephone`, to a richness of its phonological form, that distinguishes this word more than others, the converse holds for `plug`.

### *6.1.4 Non Linguistic Categories and Prototypes*

Two additional experiments have been performed with this model. The first one was aimed at comparing linguistic and pre-linguistic categorization. Representations in PFC are the result of the tension between the normativity in naming things, and the natural similarities of their appearance. In LOC only the visual features are in place to inform how to shape categories. The two representations have been compared using a conventional SOM, the algorithm introduced in Sect. 4.1.3, to cluster the output of LOC and PFC, for all object samples, in a $7 \times 7$ space, sufficient for allocating the 38 categories.

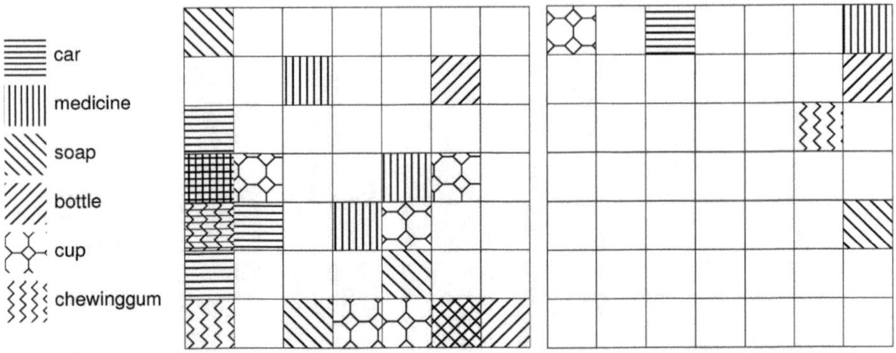


**Fig. 6.7** Comparison between pre-linguistic categorization in LOC and linguistic categorization in PFC, for some classes of objects

In Fig. 6.7 some classes of objects are compared in the SOM clustering LOC results and the other clustering PFC results. It is not too surprising that in the case of PFC there is a close mapping between SOM elements and linguistic classes of objects, while in the case of LOC they are spread along multiple SOM elements, with several overlaps. In some cases the LOC "natural" categorization includes overlaps of linguistic classes, as shown in Fig. 6.8. We can see how the two cells (far left column) where the linguistic `car` category is spread share objects of categories `chewing gum` (upper cell) and `medicine`, `cigarettes` (lower cell), likely due to the similar color contrast in the labels, in yellow and light green. The sharing of categories `soap` e `bottle` (bottom row) is probably also due to the similarity in their labels. More surprising is the presence of `kitten`, possibly by virtue of the dominant white background and the upright shape.

The additional experiments aimed at simulating types of mental imagery, a context in which the name corresponding to a known category of objects is uttered, without any relevant extant visual content. The analysis retrieves which object image is activated by area PFC. In mathematical terms, the model is presented with an input of type $c = \langle \epsilon, \breve{n} \rangle$, with $\breve{n}$ the sound corresponding to a given category name. The image $\hat{o}$ most likely elicited by this event is computed by the following equation:

$$\hat{o} = \underset{o_i \in \bigcup O,\, n_{o_i} \in N_{o_i}}{\arg\min} \left\{ \sum_{j=1 \cdots M} \alpha^j \left( x^{\mathrm{PFC}}_{p(\langle \epsilon, \breve{n} \rangle)_j}(\langle \epsilon, \breve{n} \rangle) - x^{\mathrm{PFC}}_{p(\langle \epsilon, \breve{n} \rangle)_j}(\langle o_i, n_{o_i} \rangle) \right)^2 \right\}, \qquad (6.18)$$

where $p(S)_j$ denotes the $j$-th element in the ordered set $p(S)$, as in (4.19), the function $x^{\mathrm{PFC}}(\cdot)$ is that defined by (6.17), and $N_{o_i}$ is the set of sounds naming the category of object $o_i$.

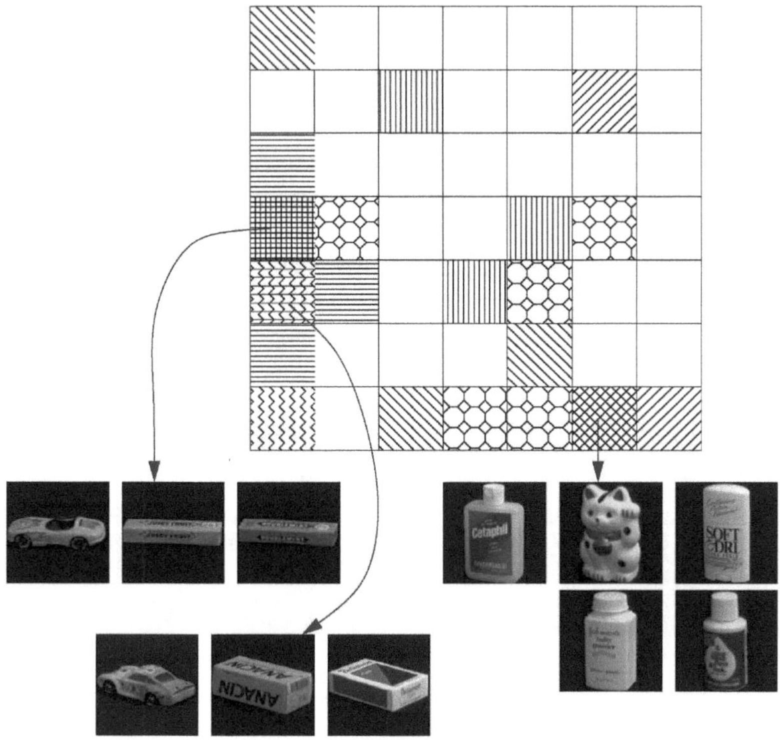

**Fig. 6.8** Examples of objects in the pre-linguistic categories of LOC map, which overlap with linguistic categories in PFC

Several examples of images elicited in such way are shown in Fig. 6.9. We can see that often all possible voices elicit the same view of an object, which can be held to represent a sort of prototype for its class, being – as in the semantic theory of prototypes (see Sect. 5.2.2) – the visual content most representative of a category. As in the experiments of pre-linguistic categorization, there are cases in which misunderstandings take place, like for `kitten`, and probably due to the reasons mentioned earlier. The confusion for `dog` is probably purely phonological, due to a close similarity with the spectrogram of `jug`. Let us mention a further investigation that may better clarify the "imagery" possibility in the model. The function evaluated for minimization in (6.18) can be used as a measure on how reliably the selected image is associated with the word. Furthermore, instead of the minimum, its histogram with respect to objects can be analyzed, in order to assess to what extent an object prevails over others. This direction of analysis has not been undertaken yet.

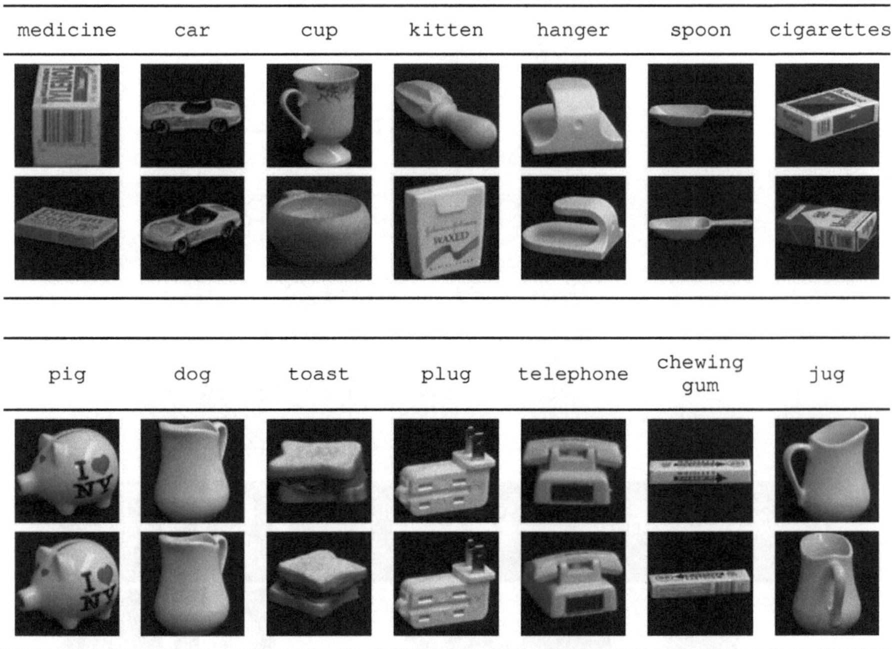


**Fig. 6.9** Examples of images elicited in PFC by voices of a category name, female voices in the *first row*, male voices in the *second row*

## 6.2 Early Lexicon Building

This model uses components of the previous one, and the overall architecture is similar, but with different purposes: the simulation of specific known psycholinguistics phenomena that are typical of lexical acquisition.

The transition from a pre-linguistic phase to a linguistic one is an extremely important moment for scrutinizing the mechanisms at the basis of the construction of language in the mind. Therefore, it is imperative to address data coming from developmental psycholinguistics and compare it with the outcomes of a brain plausible simulation.

Several particular characteristics have been consistently observed in child development, some refer to the peculiar trend in the speed with which words are learned. The most commonly held view in the literature has referred to this rapid pace in the increase of the child's vocabulary at around 18 months of age, as the "vocabulary spurt" (Lifter and Bloom 1989; Plunkett 1993). However, it has been recast over the years as being more of a gradual linear increase in the child's vocabulary development rather than an "explosion", that can be attributed to a number of factors, such as the child's increasing experience with language as well as to the development of a widening range of cognitive abilities (Elman et al. 1996; Bloom 2000; Ganger and Brent 2004).

A second phenomena is what is known as "fast mapping" (Carey 1978; Dickinson 1984), the ability to grasp aspects of the meaning of a new word on the basis of only a few incidental exposures, gained at around 2 years of age. There is a correlation between the two phenomena, in that fast mapping takes place easier once the threshold of the vocabulary spurt has been reached.

Several of the hypotheses proposed to explain these two phenomena, invoke a shift toward mental processes other than those involving development driven by sensorial experience. An idea that spread widely in psychology is rooted in the philosophical notion of "natural kind" thanks to Kripke (1972) and Putnam (1975). In the philosophical literature the intent is to separate ontology (category individuation) and conceptualization, claiming that when we name natural kinds (animals, chemical substances etc.) we want our names to refer to the very essence of those things, irrespective of the contingent representations we have formed of them. While in philosophy this claim regards what the words refer to, in psychology it has been pushed a step further, in that categories are not just based on a simple evaluation of perceptual similarity, but it is argued that there must be a deeper theoretical core, based mainly on causal relationships, that structure the way we categorize objects (Murphy and Medin 1985; Carey and Spelke 1996). In a later stage of development children are able to detect hidden causal powers, which would systematically lead the process of category formation. There is a large debate on the issue, for example Mandler (2004) denies the need for innate core knowledge, and suggests a distinction between perceptual and "conceptual" learning, but Eimas and Quinn (1994) are critical of this distinction. Tomasello (1999, 2003) points in a different direction, invoking, in addition to inductive generalization of linguistic categories, the sensitivity to social cues, like eye gaze, which help the child in understanding the intended reference of the speaker. A comparison of all the different positions is beyond the current scope, but a deeper discussion is in (Plebe et al. 2010), the purpose of the model here presented is to try to assess to what extent sensorial experiences alone can account for the two phenomena at an early stage of development.

At the risk of being repetitive, we would like to reaffirm here, how the explanation behind our proposed model hinges on the coincidence detection mechanism, whose application to word learning has been detailed in Sect. 6.1. An interesting and similar approach has been recently used by (Mayor and Plunkett 2010), with a model based on two SOM maps (see Sect. 4.1.3), one for the visual input and the other for the acoustic input. The units of the two maps are connected, and their efficiencies are updated by increasing those connecting the winning units, as well as the neighboring units on each map. Implicitly, it is an implementation at the top level of the coincidence detection mechanism. The meaning of a word is coded in the cross-connections between the two maps, and is the result of detecting repeated coincidental activation of units responding to object forms and sound forms. The main difference between Mayor and Plunkett's model and ours is in its computational grounding: their model belongs to the connectionist paradigm, and departs from mapping criteria in that the components have no mapping with any particular brain part.

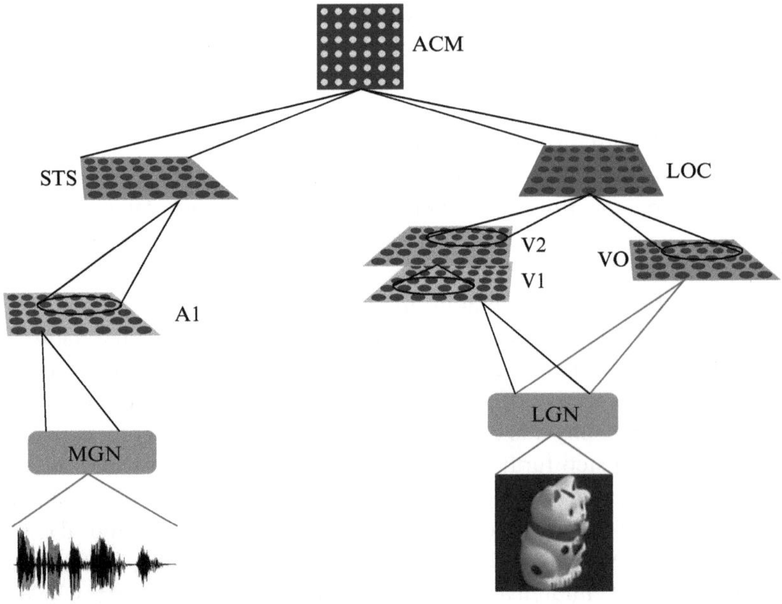


**Fig. 6.10** Scheme of the model used in experiments of lexical acquisition. All components are the same as those in the model in Fig. 6.1, except for ACM (*Abstract Categorical Map*)

The authors argue that this limitation may also be the source of difficulties encountered within their model, in particular the "inability to learn new words after the visual and auditory maps have stabilized" (Mayor and Plunkett 2010, p. 20). In their opinion, this could be overcome by employing "hierarchies (or heterarchies) of maps in both the visual and auditory pathways of the model, [so as to mimic] aspects of the organization of the visual and auditory cortex". This is precisely one virtue of our model. The overall scheme of the model is shown in Fig. 6.10, the main difference from the previous model is in the confluence between the visual and auditory streams: not in a LISSOM type map, like those seen in Sect. 4.2, but a SOM (see Sect. 4.1.3), called ACM (*Abstract Categorical Map*). Note that in the model of Mayor and Plunkett SOM maps were also used. The main difference is in their placing: in their model the SOM maps act as a substitute to a detailed hierarchy of visual and auditory cortical processes. In our model the hierarchy is preserved as much as possible, and the SOM is used as an abstract placeholder of the top level mechanism only, which is spread too widely in the brain to be realistically reproduced, as explained in Sect. 6.1.1.

Calling $\mathbf{o}_{LOC}$ and $\mathbf{o}_{STS}$ the entire content of maps LOC and STS, the input vectors to ACM are composed as follows:

$$\mathbf{v}_{ACM} = \begin{bmatrix} \mathbf{o}_{STS} \\ \mathbf{o}_{LOC} \end{bmatrix}. \tag{6.19}$$

All such vectors are presented to the ACM map according to Eq. (4.4), and the map develops using the rule (4.5).

**Table 6.5** The five stages of development, with the corresponding number of objects and words known by the model

| Stage | # objects | # words |
|---|---|---|
| Stage I | 30 | 20.6 |
| Stage II | 50 | 28.3 |
| Stage III | 60 | 31.1 |
| Stage IV | 70 | 33.5 |
| Stage V | 80 | 35.5 |

### *6.2.1 Learning Stages*

The experiments done with the previous model, in Sect. 6.1, involved a single artificial "subject". For the purposes of this model, the experiment is carried out on 500 "subjects", which are clones of the same basic model, but with different experiences. Five stages of lexical learning have been simulated, and in each stage 100 model subjects have been recruited. There is not much difference in the type of experience the models are exposed to, with respect to that visible in Fig. 6.2, the only additions are the neutral combinations of sounds and images, without semantic content, useful to the balancing of the overall number of stimulations. The visual neutral scene is a random image of the *Flowers and Landscape* McGill collection (http://www.tabby.vision.mcgill.ca/), the auditory neutral pattern is a random fragment from Wagner, *Der Fliegende Hollander*.

The 500 artificial subject are individualized from the common model, by generating 500 different subsets from the 100 COIL objects, to which the copies of the model are exposed during the linguistic development phase. The extracted subsets of objects are grouped into five different sizes, corresponding to the five linguistic stages, each with 100 subsets. Although each stage has a fixed number of known objects, because categories of objects in the COIL collection have an uneven number of exemplars, and being that the set of stimuli is selected randomly, the number of known words in a single stage of development varies slightly between individual models. In Table 6.5 the number of objects at each stage are shown, and the average number of words.

All subsets lack an entire category of objects, as well as a small number of other objects, in order to have samples that are unknown to every artificial subject, to be used as exemplars in the triadic trials. The `car` category has been chosen because it is composed by a sufficient number of different samples, seven. The use of semantically void samples, aggregating music and landscape images, allows the use of a uniform number of samples for all artificial subjects, avoiding possible artifacts in the results due to the size of the sample set.

### *6.2.2 Lexical Organization in ACM*

Since the amodal map in this model is a SOM, it is not possible to apply population coding analysis (see Sect. 4.3), in interpreting its semantic organization, as done for the PFC map in the previous model. For the SOM a simple labeling operation suffices, in accordance with its *winner-take-all* principle (see Sect. 4.1.3). Being $o$ an object of the COIL set $\mathscr{O}$, $\mathscr{W}$ the set of names of categories, the $i$-th unit of the SOM may assume one of the three labels given by the following functions:

$$l^{(I,c)}(i) = \arg\max_{o\in\mathscr{O}} \left\{\left|\left\{I_j^{(o)} : i = v\left(I_j^{(o)}, c(o)\right)\right\}\right|\right\}, \tag{6.20}$$

$$l^{(c|I)}(i) = \arg\max_{c\in\mathscr{W}} \left\{\left|\left\{I_j^{(o)} : c = c(o) \wedge i = v\left(I_j^{(o)}, c\right)\right\}\right|\right\}, \tag{6.21}$$

$$l^{(I)}(i) = \arg\max_{o\in\mathscr{O}} \left\{\left|\left\{I_j^{(o)} : i = v\left(I_j^{(o)}, \epsilon\right)\right\}\right|\right\}, \tag{6.22}$$

where $I_j^{(o)}$ is an image of the COIL database representing object $o$ at viewpoint $j$, $c(o) : \mathscr{O} \to \mathscr{W}$ is the function giving the lexical category of the object $o$, and $v(\cdot,\cdot)$ the function associating an image and a word given as input to the model with a winner neuron in the SOM.

We maintain the convention of calling $\epsilon$ a null argument in $v(\cdot,\cdot)$, and $|\cdot|$ in this context is the cardinality of a set. The function $l^{(I,c)}(\cdot)$ in Eq. (6.20) identifies a category by joint visual aspect and listening to its name, the object identification by vision only is $l^{(I)}(\cdot)$ in Eq. (6.22), labeling of recognized categories of objects is given by function $l^{(c,I)}(\cdot)$ in Eq. (6.21).

Applying the labeling functions, it is now possible to verify the correctness of the identification or the categorization of an object presented to the model, by checking the label of the winner neuron in the ACM map. For example, the performance of the model when presented with an object and its name are evaluated using the labeling from Eq. (6.20). If the winner neuron has the same label of the object, the model performed correctly. When presenting an object without naming it, the procedure is the same, but using as labeling function Eq. (6.22). The overall accuracy is given by the fraction of correct judgments, for example in the case (6.20), recognition of objects by visual aspect and category, the accuracy is:

$$a^{(I,c)}(o) = \frac{\left|\left\{I_i^{(o)} : l\left(v\left(I_i^{(o)}, c(o)\right)\right) = o\right\}\right|}{\left|\left\{I_i^{(o)}\right\}\right|}. \tag{6.23}$$

While functions $l^{(\cdot)}(\cdot)$ maps from set of objects and/or names to neurons in ACM, it is also easy to compute the inverse, for example as in the following function:

$$m^{(I,c)}(o) = \arg\max_{i\in\{1,\ldots,M\}} \left\{\left|\left\{I_i^{(o)} : i = v\left(I_i^{(o)}, c(o)\right)\right\}\right|\right\}, \tag{6.24}$$

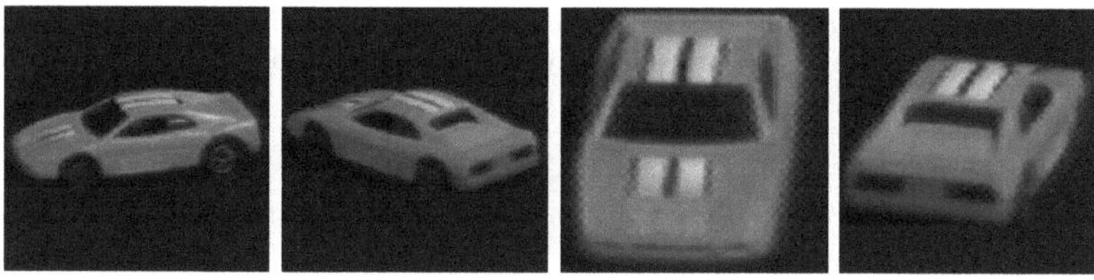

**Fig. 6.11** The standard exemplar used in fast-mapping experiments, object `car`, #23 of the COIL-100 collection

maps from object $o$ into a neuron $i$ over all $M$ neurons in the ACM SOM map, the neuron that is the most likely to win on that object, when its category $c(o)$, is also named.

### 6.2.3 Experiments of Fast-Mapping

The experiments presented here simulate the protocol typically used in psycholinguistic experiments for assessing fast-mapping. In the training stage an unknown object is presented to the child, labeled with a non-existing word, the typical utterance is *this is a DAX*. In the test stage a small set of objects is presented, and the child is asked to identify the new one: *give me a DAX* (Smith 2001; Regier 2005).

The standard used in the experiment is object #23 of the COIL-100 collection, displayed in Fig. 6.11. During the training stage it is shown to the model under 3 different views, very briefly, for 40 training epochs only, as in the fast-mapping experiment with children.

In the next stage, that of testing, how do we go about making the model "point" to the chosen object? Mathematically, of course. The ACM map is not evaluated for the winner neuron, as in the ordinary SOM equation (4.4). The responses of both new objects at the target unit of the standard object #23, given by Eq. (6.20), are compared. Let $i_{\#23} = m^{(I,c)}(o_{\#23})$ be this target unit, $\breve{o}$ the test object of the same category, and $\tilde{o}$ the test object of a different category. Let us call $\mathbf{v}(o)$ the input to ACM, as given by Eq. (6.19), in response to the presentation of an object $o$ in its frontal view, when the corresponding category is named. The model makes the correct choice if the following inequality holds:

$$\|\mathbf{v}(\breve{o}) - \mathbf{x}_{i_{\#23}}\| < \|\mathbf{v}(\tilde{o}) - \mathbf{x}_{i_{\#23}}\| \qquad (6.25)$$

For most of the objects in COIL-100, taken in turn as the strange object in the tests, all 500 artificial subjects made the correct choice, therefore the analysis has been made by concentrating exclusively on the few objects that confounded the models, the results are reported in Table 6.6.

**Table 6.6** Results of the fast-mapping experiments. Numbers are a fraction of correct responses when presenting the object of the same category (*column*) and an object of a different category (*row*)

| | |  |  | 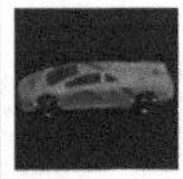 |  | 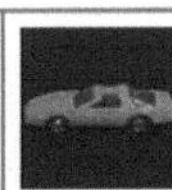 |  |
|---|---|---|---|---|---|---|---|
| | | #6 | #8 | #15 | #69 | #76 | #91 |
| #38 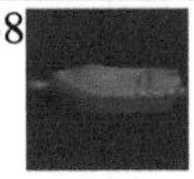 | Stage I | 0.25 | 0.23 | 0.58 | 0.25 | 0.96 | 0.69 |
| | Stage II | 0.18 | 0.17 | 0.58 | 0.19 | 0.93 | 0.67 |
| | Stage III | 0.14 | 0.14 | 0.63 | 0.18 | 0.94 | 0.71 |
| | Stage IV | 0.26 | 0.24 | 0.76 | 0.27 | 0.96 | 0.81 |
| | Stage V | 0.25 | 0.25 | 0.90 | 0.26 | 0.92 | 0.91 |
| #46 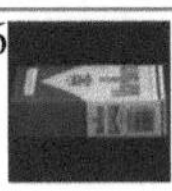 | Stage I | 0.78 | 0.87 | 1.00 | 0.87 | 1.00 | 1.00 |
| | Stage II | 0.84 | 0.92 | 1.00 | 0.93 | 1.00 | 1.00 |
| | Stage III | 0.80 | 0.91 | 1.00 | 0.91 | 1.00 | 1.00 |
| | Stage IV | 0.92 | 0.95 | 1.00 | 0.95 | 1.00 | 1.00 |
| | Stage V | 0.94 | 0.96 | 1.00 | 0.98 | 1.00 | 1.00 |
| #47 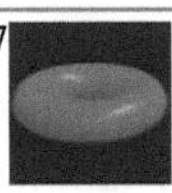 | Stage I | 0.60 | 0.63 | 0.88 | 0.64 | 0.98 | 0.90 |
| | Stage II | 0.63 | 0.65 | 0.92 | 0.67 | 0.96 | 0.93 |
| | Stage III | 0.63 | 0.70 | 0.93 | 0.69 | 0.97 | 0.96 |
| | Stage IV | 0.80 | 0.83 | 0.96 | 0.81 | 1.00 | 0.97 |
| | Stage V | 0.84 | 0.84 | 0.98 | 0.85 | 1.00 | 0.99 |
| #100  | Stage I | 0.35 | 0.39 | 0.79 | 0.35 | 0.96 | 0.87 |
| | Stage II | 0.41 | 0.41 | 0.82 | 0.34 | 0.97 | 0.88 |
| | Stage III | 0.40 | 0.45 | 0.85 | 0.37 | 0.98 | 0.90 |
| | Stage IV | 0.54 | 0.55 | 0.91 | 0.41 | 1.00 | 0.94 |
| | Stage V | 0.67 | 0.67 | 0.91 | 0.43 | 1.00 | 0.95 |

It is immediately apparent that all models attested fast-mapping capabilities, as well as fast categorization: the models did not have a previous category `car`, they rapidly acquired it, and the new name behaved as glue for connecting coherent perceptual features of other new objects to this category. The only four objects that sometimes confounded the model are, at least partially, red. However, we can quickly rule out the hypothesis of color being the dominant feature in establishing a new category, since there are 13 more red objects in the COIL-100 collection that were never chosen as the standard by the model. The object that induced the most errors is #38, of category `boat`, its shape is quite similar to that of the target, especially under certain views. On the contrary, in the case of object #100, of category `truck`, the shape is quite different, but there is a structural similarity, it has similar components, such as the wheels, and is in a similar position with respect to the main body. The choice of this wrong object, especially during the first age groups, may indicate an initial tendency to categorize by taking into account object components, as in Biederman (1987).

An important result of the experiments is the progressive improvement in the ability to grasp new categories, at later stages of development. This is far from being a trivial fact for a model, for which "development" just means the dimension of its vocabulary, without any difference in its neural architecture. This trend can be appreciated in its graphic form in Fig. 6.12, and with a statistical assessment given in Table 6.7. Further details of these experiments are in Plebe et al. (2010).

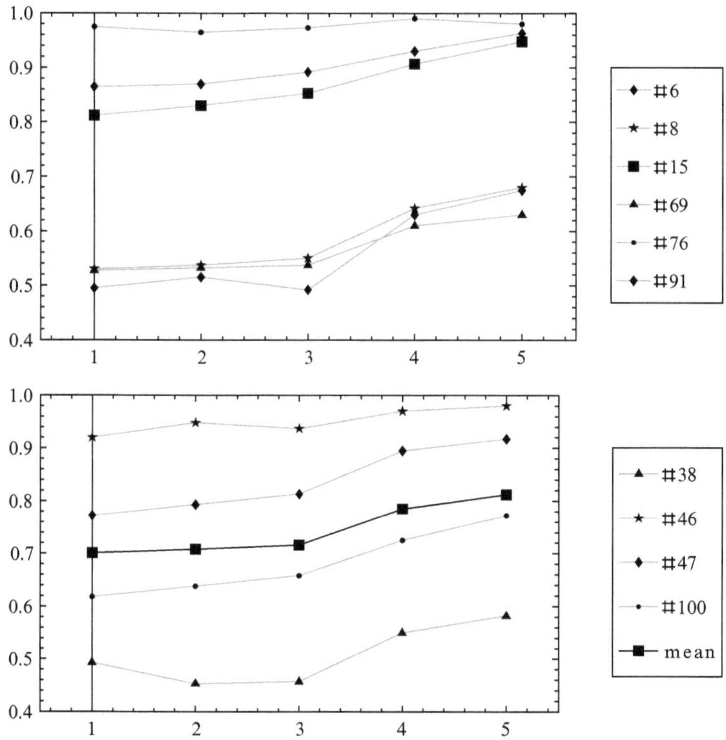


**Fig. 6.12** Success in fast-mapping episodes, at different stages of development. In the *top plot* curves of success for each object of the same category of the standard `car`, in the *bottom plot* the curves for each object of category other than the standard, and the mean trend

**Table 6.7** Statistical analysis of the progress in fast-mapping with increasing stages of development

| Groups | $f_{(\cdot,\cdot)}$ | $p$ |
|---|---|---|
| Car objects | $f_{(4,20)} = 20.0$ | $< 0.001$ |
| Non-car objects | $f_{(4,12)} = 16.4$ | $< 0.001$ |

## References


Ashby, F. G., & Spiering, B. J. (2004). The neurobiology of category learning. *Behavioral and Cognitive Neuroscience Reviews, 3*, 101–113.

Bates, E., Dal, P. S., & Thal, D. (1995). Individual differences and their implications for theories of language development. In P. Fletcher & B. M. Whinney (Eds.), *Handbook of child language* (pp. 96–151). Oxford: Basil Blackwell.

Bednar, J. A., & Miikkulainen, R. (2004). Prenatal and postnatal development of laterally connected orientation maps. *Neurocomputing, 58*, 985–992.

Biederman, I. (1987). Recognition-by-components: A theory of human image understanding. *Psychological Review, 94*, 115–147.

Black, A. W., & Taylor, P. A. (1997). The festival speech synthesis system: System documentation. Technical report HCRC/TR-83, Human Communication Research Centre, University of Edinburgh, Edinburgh.

Bloom, P. (2000). *How children learn the meanings of words*. Cambridge: MIT.

Bornstein, M. H., & RCote, L. (2004). Cross-linguistic analysis of vocabulary in young children: Spanish, Dutch, French, Hebrew, Italian, Korean, and American english. *Child Development, 75*, 1115–1139.

Brown, M. C. (2003). Audition. In L. R. Squire, F. Bloom, S. McConnell, J. Roberts, N. Spitzer, & M. Zigmond (Eds.), *Fundamental neuroscience* (pp. 699–726). New York: Academic.

Carey, S. (1978). The child as word learner. In M. Halle, J. Bresnan, & G. Miller (Eds.), *Linguistic theory and psychological reality* (pp. 264–293). Cambridge: MIT.

Carey, S., & Spelke, E. (1996). Science and core knowledge. *Journal of Philosophy of Science, 63*, 515–533.

Chapman, B., Stryker, M. P., & Bonhoeffer, T. (1996). Development of orientation preference maps in ferret primary visual cortex. *Journal of Neuroscience, 16*, 6443–6453.

Deco, G., & Rolls, E. (2004). A neurodynamical cortical model of visual attention and invariant object recognition. *Vision Research, 44*, 621–642.

Dickinson, D. K. (1984). First impressions: Children's knowledge of words gained from a single exposure. *Applied Psycholinguistics, 5*, 359–373.

Dowling, J. E. (1987). *The retina: An approachable part of the brain*. Cambridge: Cambridge University Press.

Edelman, S., & Duvdevani-Bar, S. (1997). A model of visual recognition and categorization. *Philosophical Transactions of the Royal Society of London, 352*, 1191–1202.

Eimas, P. D., & Quinn, P. C. (1994). Studies on the formation of perceptually based basic-level categories in young infants. *Child Development, 3*, 903–917.

Elman, J. L., Bates, E., Johnson, M. H., Karmiloff-Smith, A., Parisi, D., & Plunkett, K. (1996). *Rethinking innateness a connectionist perspective on development*. Cambridge: MIT.

Fairhall, S. L., & Caramazza, A. (2013). Brain regions that represent amodal conceptual knowledge. *Journal of Neuroscience, 33*, 10,552–10,558.

Farah, M. J. (1990). *Visual agnosia: Disorders of object recognition and what they tell us about normal vision*. Cambridge: MIT.

Freedman, D. J., Riesenhuber, M., Poggio, T., & Miller, E. K. (2001). Categorical representation of visual stimuli in the primate prefrontal cortex. *Science, 291*, 312–316.

Freedman, D. J., Riesenhuber, M., Poggio, T., & Miller, E. K. (2003). Visual categorization and the primate prefrontal cortex: Neurophysiology and behavior. *Journal of Neurophysiology, 88*, 929–941.

Fuster, J. M. (2008). *The prefrontal cortex* (4th ed.). New York: Academic.

Ganger, J., & Brent, M. R. (2004). Reexamining the vocabulary spurt. *Developmental Psychology, 40*, 621–632.

Gershkoff-Stowe, L., & Smith, L. B. (2004). Shape and the first hundred nouns. *Child Development, 75*, 1098–1114.

Gödecke, I., & Bonhoeffer, T. (1996). Development of identical orientation maps for two eyes without common visual experience. *Nature, 379*, 251–254.

Grassman, S., Stracke, M., & Tomasello, M. (2009). Two year olds exclude novel objects as potential referents of novel words based on pragmatics. *Cognition, 112*, 488–493.

Huey, E. D., Krueger, F., & Grafman, J. (2006). Representations in the human prefrontal cortex. *Current Directions in Psychological Science, 15*, 167–171.

Kashimori, Y., Ichinose, Y., & Fujita, K. (2007). A functional role of interaction between IT cortex and PF cortex in visual categorization task. *Neurocomputing, 70*, 1813–1818.

Katz, L., & Shatz, C. (1996). Synaptic activity and the construction of cortical circuits. *Science, 274*, 1133–1138.

Khaligh-Razavi, S. M., & Kriegeskorte, N. (2014). Deep supervised, but not unsupervised, models may explain it cortical representation. *PLoS Computational Biology, 10*, e1003,915.

Kripke, S. A. (1972). Naming and necessity. In D. Davidson & G. H. Harman (Eds.), *Semantics of natural language* (pp. 253–355). Dordrecht: Reidel Publishing.

Lifter, K., & Bloom, L. (1989). Object knowledge and the emergence of language. *Infant Behavior and Development, 12*, 395–423.

Mandler, J. M. (2004). *The foundations of mind*. Oxford: Oxford University Press.

Mastronarde, D. N. (1983). Correlated firing of retinal ganglion cells: I. Spontaneously active inputs in X- and Y-cells. *Journal of Neuroscience, 14*, 409–441.

Mayor, J., & Plunkett, K. (2010). A neurocomputational account of taxonomic responding and fast mapping in early word learning. *Psychological Review, 117*, 1–31.

Miller, E. K., Freedman, D. J., & Wallis, J. D. (2002). The prefrontal cortex: Categories, concepts and cognition. *Philosophical Transactions: Biological Sciences, 357*, 1123–1136.

Murphy, G. L., & Medin, D. L. (1985). The role of theories in conceptual coherence. *Psychological Review, 92*, 289–316.

Näger, C., Storck, J., & Deco, G. (2002). Speech recognition with spiking neurons and dynamic synapses: A model motivated by the human auditory pathway. *Neurocomputing, 44–46*, 937–942.

Nayar, S., & Murase, H. (1995). Visual learning and recognition of 3-d object by appearance. *International Journal of Computer Vision, 14*, 5–24.

Pereira, A. F., Smith, L. B., & Yu, C. (2014). A bottom-up view of toddler word learning. *Psychon Bulletin Review, 21*, 178–185.

Plebe, A. (2007a). A model of angle selectivity development in visual area V2. *Neurocomputing, 70*, 2060–2066.

Plebe, A. (2007b). A neural model of object naming. *Enformatika, 2*, 130–135.

Plebe, A. (2012). A model of the response of visual area V2 to combinations of orientations. *Network: Computation in Neural Systems, 23*, 105–122.

Plebe, A., Domenella, R. G. (2005). The emergence of visual object recognition. In W. Duch, J. Kacprzyk, E. Oja, S. Zadrony (Eds.), *Artificial Neural Networks – ICANN 2005 15th International Conference*, Warsaw (pp. 507–512). Berlin: Springer.

Plebe, A., Domenella, R. G. (2006). Early development of visual recognition. *BioSystems, 86*, 63–74.

Plebe, A., Domenella, R. G. (2007). Object recognition by artificial cortical maps. *Neural Networks, 20*, 763–780.

Plebe, A., De La Cruz, V. M., & Mazzone, M. (2007). Artificial learners of objects and names. In Y. Demiris, B. Scassellati, & D. Mareschal (Eds.), *Proceedings of the 6th International Conference on Development and Learning, IEEE* (pp. 300–305). London: Imperial College.

Plebe, A., Mazzone, M., & De La Cruz, V. M. (2010). First words learning: A cortical model. *Cognitive Computation, 2*, 217–229.

Plebe, A., Mazzone, M., & De La Cruz, V. M. (2011). A biologically inspired neural model of vision-language integration. *Neural Network World, 21*, 227–249.

Plunkett, K. (1993). Lexical segmentation and vocabulary growth in early language acquisition. *Journal of Child Language, 20*, 43–60.

Putnam, H. (1975). The meaning of "meaning". In H. Putnam, *Mind, language and reality* (Vol. 2, pp. 215–271). Cambridge: MIT.

Regier, T. (2005). The emergence of words: Attentional learning in form and meaning. *Cognitive Science, 29*, 819–865.

Riesenhuber, M., & Poggio, T. (2000). Models of object recognition. *Nature Neuroscience, 3*, 1199–1204.

Rogers, T. T., & McClelland, J. L. (2006). *Semantic cognition – A parallel distributed processing approach*. Cambridge: MIT.

Rolls, E. T., & Stringer, S. M. (2006). Invariant visual object recognition: A model, with lighting invariance. *Journal of Physiology – Paris, 100*, 43–62.

Rumelhart, D. E., & McClelland, J. L. (Eds.) (1986). *Parallel distributed processing: Explorations in the microstructure of cognition*. Cambridge: MIT.

Smith, L. B. (2001). How domain-general processes may create domain-specific biases. In M. Bowerman & S. Levinson (Eds.), *Language acquisition and conceptual development*. Cambridge: Cambridge University Press.

Taylor, N. R., Hartley, M., & Taylor, J. G. (2005). Coding of objects in low-level visual cortical areas. In W. Duch, J. Kacprzyk, E. Oja, & S. Zadrony (Eds.), *15th international conference proceedings artificial neural networks (ICANN '05)* (pp. 57–63). Berlin: Springer.

Thompson, I. (1997). Cortical development: A role for spontaneous activity? *Current Biology, 7*, 324–326.

Tomasello, M. (1999). *The cultural origins of human cognition*. Cambridge: Harvard University Press.

Tomasello, M. (2003). *Constructing a language: A usage-based theory of language acquisition*. Cambridge: Harvard University Press.

Volkmer, M. (2004). A pulsed neural network model of spectro-temporal receptive fields and population coding in auditory cortex. *Neural Computing, 3*, 177–193.

Wallis, G., & Rolls, E. (1997). Invariant face and object recognition in the visual system. *Progress in Neurobiology, 51*, 167–194.

Wood, J. N., & Grafman, J. (2003). Human prefrontal cortex: Processing and representational perspectives. *Nature Reviews Neuroscience, 4*, 139–147.

Yu, C., Smith, L. B. (2012). Embodied attention and word learning by toddlers. *Cognition, 125*, 244–262.

# Chapter 7
# First Syntax, Adjectives and Colors

**Abstract** A long-standing question in language development research concerns the contrast between early word learning and the learning of adjectives. The difficulty children experience early on in the acquisition of color terms, for example, led Darwin to speculate that children are initially color blind. Adjectives, in fact, are almost entirely missing in early productive vocabularies across languages. Despite the accounts proposed to explain the phenomena the debate continues to be far from being resolved. We believe adjective learning requires the development of two basic abilities. The first one is at a syntactic level: the comprehension of utterances with adjectives entails a proto-syntactic ability in discriminating the different roles of two words, by virtue of their sequence. The second is at a semantic level, which entails understanding the predicative function of the adjective, against the more basic mapping of sounds with whole objects. Both abilities develop as a result of the synergy between learning from progressive exposure to a language rich environment and the maturation of neural structures. The "visual diet" also influences the type of adjectives learned, this being particularly pertinent to what color terms are learned.

This chapter will discuss neurocomputational models that have in part simulated neural processes behind the learning of adjectives in linguistic development; how an initial sensitivity to word-order (leading to early syntactic learning or what we call proto-syntactic ability) might develop through linguistic exposure and brain maturational processes; and how both exposure to language as well as the range of colors dominant in a particular natural environment might influence not only how color terms are learned, but how colors are perceived, a topic that is right at the center of the historic linguistic relativity debate.

## 7.1 The First Syntax

Adjectives are normally heard by children together with other linguistic items and are one of the first syntactic challenges novice language learners face. Utterances with the sequence `[Adj Noun]` are what we refer to as, "embryonic-syntax", and this pattern departs from the single sound pattern to reference scheme initially experienced by infants. This explanation fits well with developmental evidence that the learning of adjectives, while difficult at first, then gets easier, once children have acquired more knowledge about their language.



A. Plebe, V.M. De La Cruz, *Neurosemantics*,
Studies in Brain and Mind 10, DOI 10.1007/978-3-319-28552-8_7

The ability to acquire adjectives very likely also depends on the maturation of brain circuits, especially in the prefrontal cortex. Language development crucially depends on the development of an expanding working memory capacity. The emergence of syntactic processes, such as being sensitive to the order in which words appear, would depend on these enhanced memory circuits found in the temporal-parietal and prefrontal areas, known to develop slowly in ontogeny. This would account for why more complex grammatical forms are acquired later in development: they depend on an expanded memory capacity that is just not available in early infancy. In this section, we discuss evidence for this and present a neurosemantic model that provides predictions further supporting this hypothesis. Less memory is necessary for learning nouns initially, but adjective learning is made possible and subsequently easier, only once memory circuits have been enhanced.

### *7.1.1 The Difficulties of Adjectives*

In contrast to the remarkable rate at which young children learn new nouns, especially during the stage often referred to as the vocabulary spurt, simulated in Sect. 6.2, the acquisition of adjectives is sluggish and their use is prone to errors. Even color adjectives, whose meaning seems easily and unambiguously related to perceptual features, seem to be particularly challenging early on in development, so much so that Darwin (1877), himself, noting the lack of color terms used by his own child, mistakenly speculated that children are initially born color blind.

Literature on this phenomenon was reviewed by Gasser and Smith (1998), which points to three kinds of evidence. First, nouns dominate early productive vocabularies of children, while adjectives are rare or nonexistent; second, experimental studies of word learning show that the application of a novel adjective appears more slowly and is more variably determined than the application of names for things; third, there is some evidence that children are more prone to errors with adjectival rather than with nominal meanings.

There are several alternative explanations for what seems to be an advantage in learning nouns as opposed to adjectives. According to Gentner (1978) there is a purely logical explanation behind this, for nouns refer to entities, and not relations between entities. For Mintz and Gleitman (2002) the difficulty is in the predicative function of adjectives, in modifying the properties of a concept their learning is necessarily grounded on the acquisition of nouns first. For further details on the discussion see Plebe et al. (2013).

The kind of explanation we propose is instead directly derived from the general mechanism of coincidence detection (see Sect. 3.2). Often an adjective refers to a single dimension in the space of perceptual features, it is the case of color, size, and shape adjectives. For this reason the adjective hooks onto a temporal coincidence that is very weak, poorly correlated with a large number of other dimensions in perceptual feature space. It is known that children have difficulty in dealing with more than one perceptual dimension at a time. This kind of difficulty vanishes when learning names of objects, since in this case the sound of the noun of the object category is associated consistently with a rich set of perceptual features. Gradually, with the maturation of the child, mechanisms of selective attention will refine the detection of coincidence between the sound of an adjective, and the single perceptual feature it is regularly correlated with, despite variations in all other dimensions.

There is an additional important aspect of co-occurrence or coincidence, and that is that adjectives are rarely heard in isolation. They are most often heard in conjunction with other linguistic items, and therefore represent one of the first experiences with syntax, or where the young listener is called upon to deal with word order. Sequences of the type `[Adj Noun]` are constituents of what could be considered an "embryonic syntax", which departs from the scheme of a holistic relation between a sound sequence and a referenced object in the world. In sum, coincidence detection appears to be an important variable in the process of coming up with a sound trace of the trajectory found in empirical studies on adjective learning (Sandhofer and Smith 2001).

### *7.1.2 Simulation of Working Memory Maturation*

From a neurocomputational point of view, the function decoding what we refer to as embryonic syntax, as in the understanding of the sequence `[Adj Noun]`, requires a specific circuital maturation, not available in the early months of life. The prefrontal cortex is involved in this kind of semantic process, and exhibits a trend of development that matches well with the trajectory of adjective learning in children (Aboitiz et al. 2006; Fuster 2001). It is well agreed upon that the prefrontal cortex is involved in recursive connections with temperoparietal areas supporting the short-term memory theorized by Baddeley (1992) and which would also support phonological decoding. Because in time the sequence of sounds to be understood increasingly become longer, there is also the increased need to temporarily retain them in memory (Vallar and Shallice 2007). The same requirements hold, at a higher level, for syntactic units. The specialization of this model, with respect to the basic model of name semantics seen in Sect. 6.1, is in accounting for the maturation of working memory between the prefrontal cortex and the superior temporal sulcus. It will be implemented by recursively computing twice the contribution of the auditory pathway in time.

The basic version of the model, without working memory, is shown in Fig. 7.1. There is a minor variation with respect to the model seen in Fig. 6.1, in the modeling of the primary auditory cortex. Instead of a single Topographica map, it is split into two distinct maps, to account for the double population of neurons in this area (Atzori et al. 2001) (see Sect. 3.4.2). This refinement is useful in this model, since the addition of adjectives has enlarged the repertoire of sounds to be encoded phonologically.

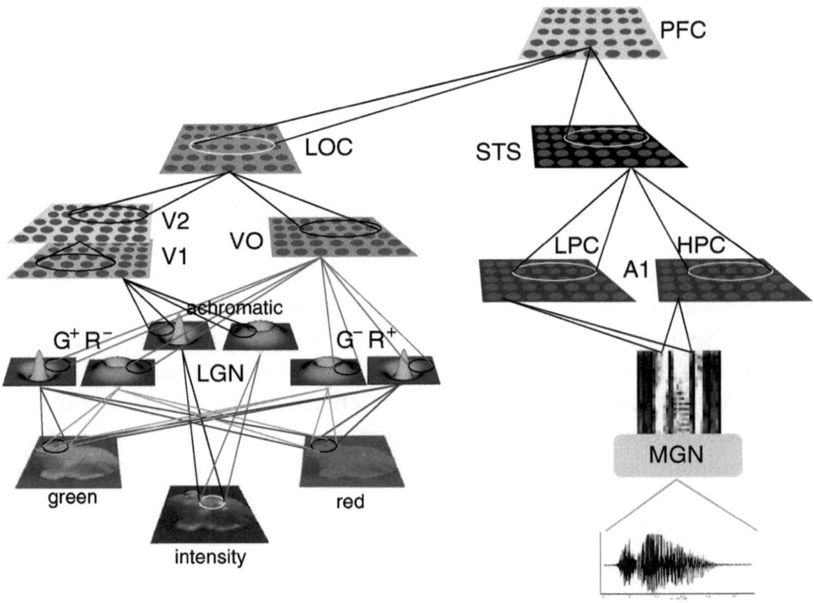


**Fig. 7.1** Basic scheme of the model without working memory. The components are the same of the model in Fig. 6.1, except for A1-LPC (Auditory Primary Cortex – Low-Probability Connections) and A1-HPC (Auditory Primary Cortex – High-Probability Connections)

Consequently, the equations of the auditory stream of processing are changed to the following:

$$x_{\tau,\omega}^{(\square)} = \left| \sum_{t=t_0}^{t_M} v(t) w(t-\tau) e^{-j\omega t} \right|^2 \tag{7.1}$$

$$\begin{aligned} x^{(\mathrm{LPC})} = f\Bigg( & \gamma_{\mathrm{A}}^{(\mathrm{LPC})} \mathbf{a}_{r_{\mathrm{A}}}^{(\mathrm{LPC}\leftarrow\square)} \cdot \mathbf{x}_{r_{\mathrm{A}}}^{(\square)} \\ & + \gamma_{\mathrm{E}}^{(\mathrm{LPC})} \mathbf{e}_{r_{\mathrm{E}}}^{(\mathrm{LPC})} \cdot \mathbf{x}_{r_{\mathrm{E}}}^{(\mathrm{LPC})} - \gamma_{\mathrm{H}}^{(\mathrm{LPC})} \mathbf{h}_{r_{\mathrm{H}}}^{(\mathrm{LPC})} \cdot \mathbf{x}_{r_{\mathrm{H}}}^{(\mathrm{LPC})} \Bigg) \end{aligned} \tag{7.2}$$

$$\begin{aligned} x^{(\mathrm{HPC})} = f\Bigg( & \gamma_{\mathrm{A}}^{(\mathrm{HPC})} \mathbf{a}_{r_{\mathrm{A}}}^{(\mathrm{HPC}\leftarrow\square)} \cdot \mathbf{x}_{r_{\mathrm{A}}}^{(\square)} \\ & + \gamma_{\mathrm{E}}^{(\mathrm{HPC})} \mathbf{e}_{r_{\mathrm{E}}}^{(\mathrm{HPC})} \cdot \mathbf{x}_{r_{\mathrm{E}}}^{(\mathrm{HPC})} - \gamma_{\mathrm{H}}^{(\mathrm{HPC})} \mathbf{h}_{r_{\mathrm{H}}}^{(\mathrm{HPC})} \cdot \mathbf{x}_{r_{\mathrm{H}}}^{(\mathrm{HPC})} \Bigg) \end{aligned} \tag{7.3}$$

$$\begin{aligned} x^{(\mathrm{STS})} = f\Bigg( & \gamma_{\mathrm{A}}^{(\mathrm{STS})} \left( \mathbf{a}_{r_{\mathrm{A}}}^{(\mathrm{STS}\leftarrow\mathrm{LPC})} \cdot \mathbf{x}_{r_{\mathrm{A}}}^{(\mathrm{LPC})} + \mathbf{a}_{r_{\mathrm{A}}}^{(\mathrm{STS}\leftarrow\mathrm{HPC})} \cdot \mathbf{x}_{r_{\mathrm{A}}}^{(\mathrm{HPC})} \right) \\ & + \gamma_{\mathrm{E}}^{(\mathrm{STS})} \mathbf{e}_{r_{\mathrm{E}}}^{(\mathrm{STS})} \cdot \mathbf{x}_{r_{\mathrm{E}}}^{(\mathrm{STS})} - \gamma_{\mathrm{H}}^{(\mathrm{STS})} \mathbf{h}_{r_{\mathrm{H}}}^{(\mathrm{STS})} \cdot \mathbf{x}_{r_{\mathrm{H}}}^{(\mathrm{STS})} \Bigg) \end{aligned} \tag{7.4}$$

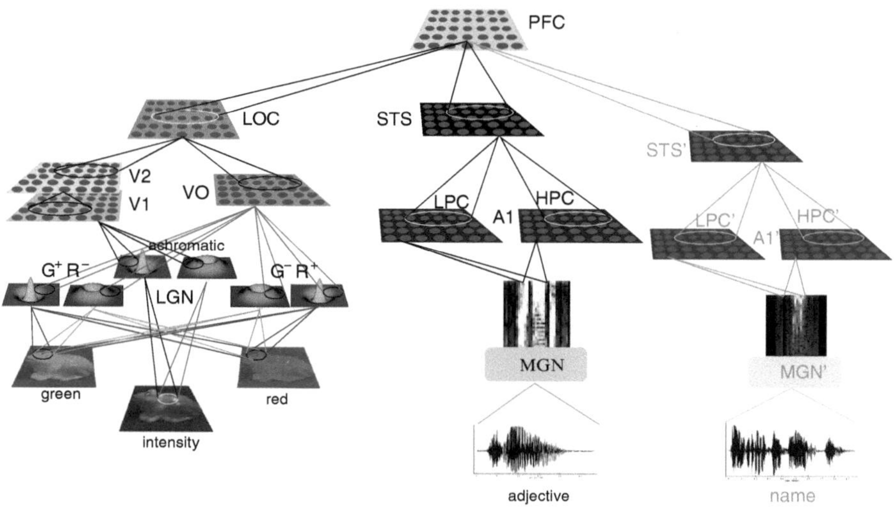


**Fig. 7.2** Complete scheme of the model, including working memory. Compared with the basic version of Fig. 7.1, there is the additional recursive connection of STS', that is STS delayed in time, to PFC

For clarity Eq. (6.7) has been rewritten here as (7.1), using the same superscript ◫ for the spectrotemporal representation of the sound waves. Equation (7.4) collects the two projections from the auditory cortex into the STS, where the phonological representations are organized.

The complete model, shown in Fig. 7.2, includes a virtual replication of STS into STS', that is a recursive contribution of the same auditory signal delayed in time, acting as working memory. The equation for PFC changes from (6.14) assuming the following form:

$$
\begin{aligned}
x^{(\mathrm{PFC}^*)} = f\Bigg( & \gamma_{\mathrm{A}}^{(\mathrm{PFC}^*)} \mathbf{a}_{r_{\mathrm{A}}}^{(\mathrm{PFC}^* \leftarrow LOC)} \cdot \mathbf{x}_{r_{\mathrm{A}}}^{(LOC)} \\
& + \gamma_{\mathrm{E}}^{(\mathrm{PFC}^*)} \mathbf{e}_{r_{\mathrm{E}}}^{(\mathrm{PFC}^*)} \cdot \mathbf{x}_{r_{\mathrm{E}}}^{(\mathrm{PFC}^*)} - \gamma_{\mathrm{H}}^{(\mathrm{PFC}^*)} \mathbf{h}_{r_{\mathrm{H}}}^{(\mathrm{PFC}^*)} \cdot \mathbf{x}_{r_{\mathrm{H}}}^{(\mathrm{PFC}^*)} \\
& + \sum_{\zeta=1}^{N_\zeta} \left( \gamma_{\mathrm{A}}^{(\mathrm{PFC}^*)} \mathbf{a}_{r_{\mathrm{A}}}^{(\mathrm{PFC}^* \leftarrow STS)} \cdot \mathbf{x}_{r_{\mathrm{A}}}^{(STS)_\zeta} \right) \Bigg)
\end{aligned}
\tag{7.5}
$$

where $\zeta$ are discrete temporal delays, corresponding to the presentation of spectrogram of progressive words in the sequence of the utterance. In this experiment $N_\zeta = 2$, since the sentence is the sequence `[Adj Noun]`. The basic form of the model is used as corresponding to early stages in development, at the onset of language acquisition, around 9–12 months of age, and the complete model at a more mature stage of development, corresponding to about 14–20 months of age.

**Table 7.1** Parameters for all neural maps of the model

| Layer | Size | $r_A$ | $r_E$ | $r_H$ | $\gamma_X$ | $\gamma_E$ | $\gamma_H$ | $\gamma_N$ |
|---|---|---|---|---|---|---|---|---|
| LGN | 112 | 2.6 | – | – | – | – | – | – |
| MGN | 32 | – | – | – | – | – | – | – |
| V1 | 96 | 8.5 | 1.5 | 7.0 | 1.5 | 1.0 | 1.0 | 0 |
| A1 | 24 | 3.5 | 2.5 | 5.5 | 5.0 | 5.0 | 6.7 | 0.8 |
| V2 | 30 | 7.5 | 8.5 | 3.5 | 50.0 | 3.2 | 2.5 | 0.7 |
| VO | 30 | 24.5 | 4.0 | 8.0 | 1.8 | 1.0 | 1.0 | 0 |
| LOC | 16 | 6.5 | 1.5 | 3.5 | 1.8 | 1.0 | 1.5 | 0 |
| STS | 16 | 3.5 | 2.5 | 2.5 | 2.0 | 1.6 | 2.0 | 0 |
| PFC | 24 | 6.5 | 4.5 | 6.5 | 1.5 | 3.5 | 4.1 | 0 |

The Table 7.1 summarizes the values of the main parameters in the equations for all the maps in the model.

### *7.1.3 Representation of Nouns and Adjectives*

The two models have been developed in a manner similar to previous models, as described in Sect. 6.1.2, running through prenatal, pre-linguistic, and a linguistic phases. During the latter, in addition to the naming of categories of objects in coincidence with their view, color adjectives are heard too. From the COIL-100 object collection (Nayar and Murase 1995), partitioned into 38 categories like in Sect. 6.1, a further partitioning has been applied with respect to color, only to objects sufficiently uniform in hue, and using the seven basic color categories. In the less mature model, the adjective is heard in isolation, in coincidence with the vision of one of the objects with the named color. In the complete model, the full sentence with the sequence `[Adj Noun]` is heard, when seeing an object of category `Noun` and color `Adj`.

The analysis of noun and adjective representation is carried out, as usual, with population coding, the methodology described in Sect. 4.3, and in particular Eq. (4.18) for establishing the population coding of a category, Eqs. (4.19) and (4.20) for evaluating the semantic performance of the model. The sets of categories in the equations have two different formulations, depending on the type of model under examination:

$$s_N^{(\mathrm{PFC})} = \langle o, n \rangle \in N = \bigcup_{O \in \mathcal{O}_N} O \times \mathcal{U}_N \tag{7.6}$$

$$s_A^{(\mathrm{PFC})} = \langle o, a \rangle \in A = \bigcup_{O \in \mathcal{O}_A} O \times \mathcal{U}_A \tag{7.7}$$

$$\begin{aligned} s_N^{(\mathrm{PFC}^*)} &= \langle o, a, n \rangle \in N \\ &= \left\{ \langle \omega, \alpha, \pi \rangle : \omega \in \mathcal{O}_N \wedge \alpha \in \mathcal{U}_{A(\omega)} \wedge \pi \in \mathcal{U}_N \right\} \end{aligned} \tag{7.8}$$

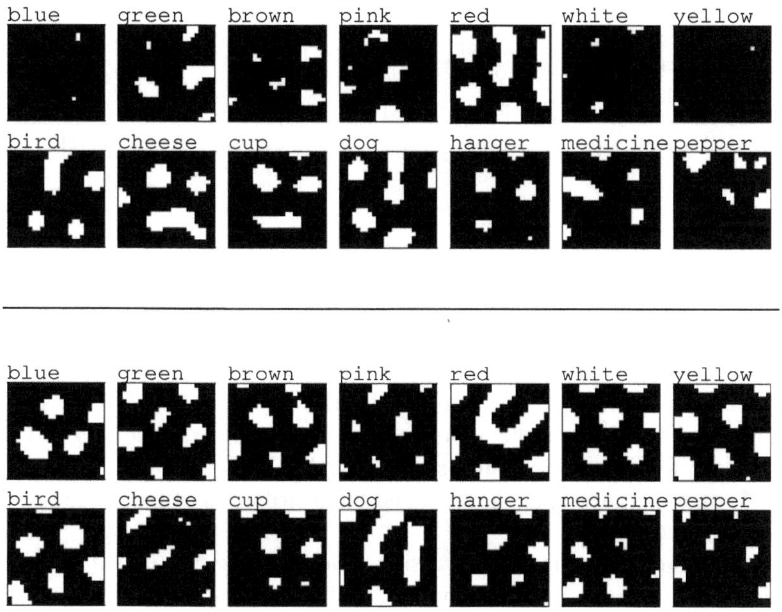


**Fig. 7.3** Comparison between population coding of object names and adjectives in the semantic map of the model. In the *top rows* PCF without working memory, in the *bottom rows* PFC* with working memory

$$\begin{aligned} s_A^{(\mathrm{PFC}^*)} &= \langle o, a, n \rangle \in A \\ &= \left\{ \langle \omega, \alpha, \pi \rangle : \omega \in \mathscr{O}_A \ \wedge \ \alpha \in \mathscr{U}_\alpha \ \wedge \ \pi \in \mathscr{U}_{N(\omega)} \right\} \end{aligned} \tag{7.9}$$

where $\mathscr{O}_N$ is the set of all images of objects that correspond to the lexical category under the noun $N$, $\mathscr{O}_A$ is the set of objects with the property consistent with the adjective $A$, $\mathscr{U}_N$ is the set of all utterances of noun $N$, and $\mathscr{U}_{N(\cdot)}$ is the set of all utterances of the noun referring to object $\cdot$, similarly for adjective utterances. In testing the immature model by Eqs. (7.6) and (7.7), the model is presented with either a simultaneous visual appearance of an object and the utterance of its name, or the simultaneous visual appearance of the object and the utterance of its adjective. Tests of the mature model by Eqs. (7.8) and (7.9) require the simultaneous presentation of a visual object and its noun utterance, followed by the delayed adjective utterance. In the first case the class collects all objects and possible adjectives pertaining to a single noun, while in the second case the class collects all objects and names pertaining to a single adjective. As an alternative to Eq. (7.9), stimuli of the form $\langle o, n, a \rangle$ will be used to test ungrammatical sentences `[Noun Adj]*`.

Figure 7.3 shows several cases of population coding for both nouns and adjectives, in the PFC and in the PFC* maps of the two models. In the case of nouns the spread and the amount of coding neurons is similar for the two models.

**Table 7.2** Accuracy of the two models in discriminating adjectives, in the case of the model with working memory the accuracy for ungrammatical sentences is tested as well

| Color | PFC | PFC* | |
|---|---|---|---|
| | | `[Adj Noun]` | `[Noun Adj]`* |
| Yellow | 0.269 | 0.845 | 0.241 |
| Red | 0.518 | 0.789 | 0.743 |
| Green | 0.297 | 0.988 | 0.671 |
| White | 0.184 | 0.922 | 0.893 |
| Brown | 0.378 | 0.997 | 0.678 |
| Pink | 0.528 | 0.789 | 0.853 |
| Blue | 0.246 | 1.000 | 0.863 |
| Mean | $0.368 \pm 0.19$ | $0.903 \pm 0.081$ | $0.715 \pm 0.226$ |

The situation is different in the case of adjectives, the weakness of the coding in PFC compared to PFC* can be appreciated visually. For example in `yellow` and `blue` the amount of coding neurons is tiny, and is very small also for `white`. In the PFC*, on the contrary, the amount and the distribution of the coding neuron is even for all adjectives.

To complement the visual impression given by Fig. 6.5 in a few cases, all numerical evaluations of the two models, gathered using Eq. (4.20), are reported in Table 7.2.

Both models show a good degree of recognition of color adjectives, however, there is a significant improvement when working memory is in place. In the less developed or immature model, accuracy is greater for nouns than for adjectives, which may sound odd, since there are 38 nouns as opposed to 9 adjectives, and noun categories easily cross boundaries of perceptual traits. This confirms that it is the stage of the model that hampers adjective learning with respect to nouns. It is interesting to note that presenting the ungrammatical sentence `[Noun Adj]`*, the advantage of working memory in the comprehension of adjectives is reduced by half. Therefore, the model in the PFC* version shows a syntactic selectivity, in that it responds better to sentences where words respect their roles, however, this behavior is not in the form of a norm, in that the violation of the syntax makes the adjective more difficult, but not impossible, to recognize.

### 7.1.4 Binding Perception, Nouns and Adjectives

The configuration of connections and receptive fields in the mature model are the mechanical consequences of the basic processes discussed in Chap. 3, from which the semantic link between perceptual stimuli and the external world is constructed. Moving in the anterior direction of the cortex, the connectivity to sensorial input in the maps becomes more and more indirect and vague. We carried out an investigation on the two different semantic classes, nouns and adjectives, to check whether their representations in the anterior maps differ significantly in their connection with the posterior maps, which are directly related with the perceptual stimuli.

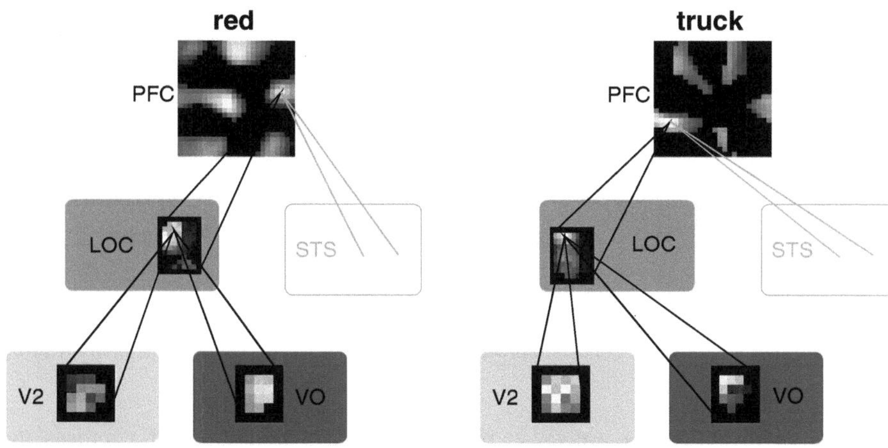


**Fig. 7.4** Examples of connectivity patterns for the noun `truck` (*right*) and the adjective `red` (*left*), tracked back from map PFC, following the receptive field of the most active coding neuron. The amount of connectivity is shown as gray level

The way to examine this is opened by the modeling strategy of segregating shape processing in V2 and color processing in VO (see Sect. 6.1.1): a discrepancy in connectivity between nouns and adjectives should be reflected in a difference with respect to those two lower areas, since objects are mainly characterized by their shape, and we used color adjectives only.

Figure 7.4 provides an example of the analysis, for a specific noun, `truck`, and an adjective `red`. Both have their own representation as population coding in the PFC map, the most active of the coding units is selected, and its connectivity tracked back to the previous map, LOC. Between all the units in LOC projecting in the receptive field of the selected PFC unit, a single one is selected again, the one with the highest synaptic efficiency. These units are again traced back, this time towards the two projecting areas: V2 and VO, each with its own receptive field. In the example of Fig. 7.4 clearly in the case of `truck` the connectivity from V2 is larger than that from VO, the other way round in the case of `red`.

While Fig. 7.4 is just a single example, the analysis has been carried over to all available nouns and adjectives in a quantitative way. For this purpose a parameter $\xi_C$ has been introduced, that measures the different amount of connections from the shape processing stream with respect to the color stream, for the population of units coding a category $C$. It is based on a preliminary parameter $\chi_C$ defined as:

$$\chi_C = \sum_{i \in \mathscr{X}_C^{(LOC)}} \frac{\Xi_\theta\left(\mathbf{a}_{r_{\text{Aa},i}}^{(\text{LOC}\leftarrow V2)}\right) - \Xi_\theta\left(\mathbf{a}_{r_{\text{Aa},i}}^{(\text{LOC}\leftarrow VO)}\right)}{\Xi_\theta\left(\mathbf{a}_{r_{\text{Aa},i}}^{(\text{LOC}\leftarrow V2)}\right) + \Xi_\theta\left(\mathbf{a}_{r_{\text{Aa},i}}^{(\text{LOC}\leftarrow VO)}\right)} \tag{7.10}$$

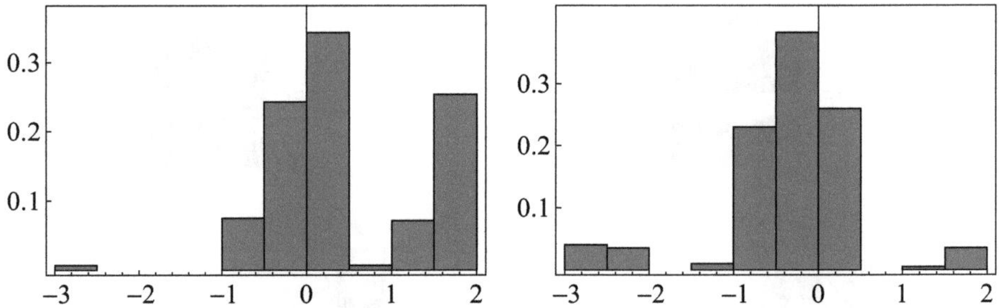


**Fig. 7.5** Distribution of $\xi$ connectivity parameter for all nouns (*left*) and all adjectives (*right*). Negative values are for larger connectivity with VO, positive values for larger connectivity with V2

where receptive fields **a** are those of Eq. (6.12), the function $\Xi_\theta(\cdot)$ returns the number of connections in the receptive field $\cdot$ whose synaptic strength is larger than $\theta$, and $\mathscr{X}_C^{(LOC)}$ is the set of units in map LOC projecting maximally into the population of units in PFC, coding for category $C$. Now the parameter $\xi$ is just the normalization of $\chi$ with respect to all the set of categories (both adjectives and nouns), and takes into account a natural discrepancy in projections from V2 and VO due to the differences in size and architecture of the two maps:

$$\xi_C = \frac{\chi_C - \overline{\chi}}{\overline{\chi}} \tag{7.11}$$

Positive values of $\xi$ indicate a pattern of connectivity that is stronger towards shape processing areas, while negative towards color processing ones. Figure 7.5 shows the distribution of $\xi$ traced back from the PFC population of neurons coding for all nouns, compared with the same distribution for all color adjectives. There is a significant difference in the distributions, in that color adjectives seem to recruit more from afferents coming from the color processing pathway than those coming from the shape processing pathway, compared to nouns. This can be interpreted as evidence of a physical grounding of the different meanings of the two linguistic classes, in the neural circuitry.

## 7.2 Color Terms and the Relativism Debate

How much of what we perceive is influenced by the terms taught to us by our linguistic community? A mechanism proposed to explain how the categories humans make may influence, or alter, how the perceived world appears is called categorical perception (CP). Research in the categorical perception of color has served as an important test bed for hypotheses on CP and for those that link language to cognition, with the domain of color terms being traditionally a privileged terrain. Color terms have been taken as evidence in favor of the linguistic relativism thesis, whose best-known formulation is the Sapir-Whorf hypothesis. The reaction of universalists has been that of searching for the underlying regularities beneath apparent lexical variation, where such regularities are mainly thought of as consequences of physiological constraints in the process of vision. Berlin and Kay (1969), in particular, proposed the well-known and influential hypothesis, according to which basic color terms follow a rigid evolutionary pattern, that is, that there would be precise rules governing how color terminology expands from the minimal repertoire of two terms to eight or more terms. Moreover, each of the languages examined would select virtually identical focal hues for the same basic colors. What this means is that putting aside minor variations, languages would differ from each other just in the number of colors they give a name to, while universal preferences would dictate the sequence of lexicalized color categories and the focal hue for each category.

Not long after Berlin and Kay published their paper on color universals, Rosch Heider and her colleagues (Heider and Olivier 1972), put their proposal to the test in a series of experiments comparing: English speaking college students with non native English speaking foreign students; speakers of a number of different languages; and English speakers with New Guinea Dani speakers (whose language was reported to possess only two basic color terms). They found evidence that focal colors were in fact, given the shortest names and remembered quicker across languages, were recognized better by both English and Dani speakers, and could be paired with names with the fewest amount of errors. They interpreted these results as being consistent with Berlin and Kay's hypothesis, including that the emergence of color lexicons followed a predetermined evolutionary course. Berlin and Kay's conclusions, however, have been called into question throughout the years from the relativist side, with a number of arguments, among which, that their findings were never objectively tested, that their assumptions and methodology led them to discard data that conflicted with their over-regularized picture, and that their data was obtained from primarily written languages and so possibly not representative of all languages (Saunders and Van Brakel 1997; Lucy 1997). In recent years, studies have attempted to reproduce Rosch Heider's results, looking for their own confirmation of the evolutionary nature of color terms. In this section we propose a model that aims to contribute to this debate, in the computational neurosemantic style. First, we will review recent findings and developments on this issue.

### 7.2.1 *Exceptions from Himba and Berinmo*

Examining linguistic relativity further, but using another method, Roberson et al. (2000, 2004, 2005), considered the question of whether two languages at the same supposed "evolutionary" stage could have similar cognitive representations of color despite having different environments. If what Kay et al. (1991) proposed, concerning the regular evolutionary pattern in which color terms would emerge in languages was correct, then this would be the case. Another aspect investigated, was the possible difference in cognitive organization between speakers of different languages, despite the similar sets of color terms. The objects of investigation were the color terms used by the Himba people of northern Namibia, a semi nomadic culturally isolated tribe, living in an arid desert-like environment, but that like the Berinmo of Papua New Guinea, have five basic color terms. Those of Himba were studied and compared with the previously studied Berinmo and with English. When recognition memory for color was examined in both Himba and Berinmo, results were consistent with Rosch Heider's results, as long as the arrays were ordered according to hue and brightness. When the arrays where randomized and the number of close competitors were instead likened to poor or best examples, neither the Himba nor the Berinmo showed memory advantages for the English best examples. What they did recognize, were the good examples of their own respective linguistic color categories, in a way that disregarded the status of these items in English color categories. A paired associate learning task (colors to pictures of familiar objects) showed the same lack of advantage for supposedly universal examples in either Berinmo or Himba speakers. Results were interpreted as showing that no single set of prototypical colors are universally cognitively privileged. The color stimuli speakers seem to remember best are those that are the best exemplars of their own named categories.

Universalist theories are further challenged by data on the acquisition of color terms, which, as remarked in Sect. 7.1.1, are particularly challenging for children to learn. Roberson et al. (2004), in fact, when comparing color naming and memory of young children learning English in the UK and children from Namibia learning Himba, found that generally, the children from both cultures appeared to acquire the color terms of their language in the same gradual manner. Results presented no indication of any advantage English-speaking children might be having in learning their color terms compared to Himba children, even though English terms map directly onto the hypothesized innate set. Furthermore, no evidence was found to indicate that children of either group had pre-partitioned representation of color at 3 years of age, before the learning of color terms. The authors argue that this evidence thus suggests that:

> if there is an innate set of cognitive categories present in young infants, then a) they are species specific and thus do not result from some property of the visual system that is shared with other primates and, b) they are not retained once adult linguistic categorization is in place.

One thing seems clear from the accumulating evidence mentioned above, learning color terms is a difficult task for children, more difficult than we might expect if what they are doing is just learning labels for innately determined universal color categories (Roberson and Hanley 2010).

### 7.2.2 Categorial Perception

Roberson et al. (2005), also investigated whether the Himba would show categorical perception at the English boundaries of green and blue, once again comparing them to speakers of Berinmo and English, and whether Himba (like the Berinmo) would show CP at boundaries within their own language that instead, do not exist in English. Subjects were shown a colored target and asked to choose, which one of two stimuli was the same as the target. Performance was facilitated in each language, when the target and the distractor had different color names (e.g. in English, a blue target with a purple distractor) as opposed to when they shared the same name (e.g. in English, two different shades of blue). All three groups showed CP, but significantly, only at the color boundaries that were clearly marked in their respective languages. Importantly, results indicated that no effect took place at the proposed universal boundary between green and blue for the Himba and Berinmo speakers, whose languages do not make this distinction. Several interesting studies that focus on a variety of aspects in color CP are worth mentioning. Anna Franklin developed methodologies for testing color perception in pre-linguistic infants, based on the oddball paradigm. Infants were first familiarized with frequent presentations of one color, and later a different color was presented. Looking time is proportional to the amount of novelty in the unfamiliar color. Interestingly, pre-linguistic infants showed categorical effects in the green/blue boundary, but in the left visual field only (Franklin et al. 2008). The results were confirmed measuring event-related potentials instead of eye movements, again for the green/blue categorical effect (Clifford et al. 2009). Franklin and her colleagues are conducting further research in order to investigate whether the change in lateralization of CP later in development, is related to the child's subsequent acquisition of color terms. This is something she does not rule out, considering there is evidence that suggests that color CP and language-mediated CP can exist alongside each other. There might even be two forms of CP: a non-lexicalized and right lateralized one in infants, and a lexicalized and left lateralized one in adults (Franklin et al. 2009).

Other experiments, have investigated the flexibility of categorical effects, and the possibility of learning new categories, inside the blue and the green regions, with just a few days of training (Özgen and Davies 2002), a result that clashes with the universality of color perception. In addition, Zhou et al. (2010) tested subjects that learned four new invented colors inside green and blue, named ang, song, duan, and ken, for lateralization, finding stronger categorical effects in the right visual field.

### 7.2.3 Tackling the Problem by Computation

While the model here presented is currently the only attempt to investigate the color relativism issue from a neurocomputational perspective, computational models of other kinds have been extensively proposed, to explore aspects of the relationship between the physical phenomenon of light and the linguistic categories of colors. One stream of research, started by Yendrikhovskij (2001) postulates that the peculiar way in which the color spectrum is split in categories by humans, simply reflects the chromatic statistics in the natural environment. He used the simple k-means clustering algorithm to cluster the color information of pixels drawn from images of natural scenes. All pixels have been projected in the $L^*u^*v^*$ CIE 1976 standard color space, using 11 clusters, their resulting position in the color space is not too far from the English color focal points. Studies on the statistical distribution of colors in natural images are highly relevant, and could indeed inform about the bias in categorization due to the natural distribution of colors in the environment. However, they clearly touch only one aspect of the matter, leaving aside all that concerns the physiology of vision and its relationship with language.

Regier et al. (2007) also used a simple abstract mathematical algorithm for partitioning the color spectra, but directly applied it to the Munsell Color Chart used in the World Color Survey. They introduced an arbitrary "well-formedness" measure of a partition in the chart, that takes into account how close together all the points in the chart under the same category are, and how far all the points of the other categories are. Using this measure as the optimizing function, they found theoretical optimal partitioning in a number of categories that looks quite similar to real partitions in selected languages with the same number of colors. The same measure has been applied to verifying that the partition in the Berinmo language turns out to be "worst", if the color chart is artificially rotated among the hue axis. For the authors, this indicates that the color naming used by Berinmo is more consistent with the universal structure of the perceptual color space, than all the other (artificial) ones. It is not clear how these results have advanced the debate since the model not only neglects any account of the physiology of vision, and its relationship with language, but also with the statistics of colors in the world. A different stream of research is trying to model communicative interactions, from which color categories are established, using artificial agents. One of the best examples of this approach is the work of Steels and Belpaeme (2005), where virtual agents engage in two types of tasks. In the discrimination game, that does not involve language, one topic has to be discriminated from several distractor colors, and the agent in isolation develops categories in order to maximize the chance that each time, the topic belongs to a category different from all the other distractors. In the guessing game, the speaker wants to get something from the listener and identifies it through language. In this task, the agent uses categories learned during the discrimination game, modifying them and developing a lexicon at the same time. Their experiments have demonstrated that linguistic interaction is able to yield a finite number of categories in a population of evolving agents. While being very interesting, particularly due to the aspect of simulated interaction, this research lacks an account of human color physiology and uses an oversimplified account of the interaction between perceptual categories and language. Recent extensions of this approach have tried to introduce elements of the human perceptual system, in a very simplified way. An example is Komarova and Jameson (2008)'s work, which simulates the presence of dichromats in a population of virtual agents.

### 7.2.4 *Brains Raised in Different Cultures*

In line with our neurosemantic approach, and differently from all the modeling efforts just reviewed, we propose a neurocomputational model of the semantics of color terms, based on the relevant brain areas, simulated with Topographica. We adapted the basic architecture of our model as already described in Sect. 7.1.1 to attempt a simplified, yet biologically plausible simulation of human color processing, and a reasonable account of the interaction between color perception and language. It is based on an earlier simpler version, developed in (Plebe et al. 2011). The scheme of the model is shown in Fig. 7.6. It has simplifications and additions with respect to the model presented in Sect. 7.1.1. It lacks areas V2 and LOC, since in this experiment it is not necessary to include shape processing. On the other side, in this model it is not possible to simplify the chromatic processing by using red, green, and intensity components, as done in all previous neurosemantic models. The precise reconstruction of human color physiology is essential for this simulation, therefore, all visual inputs have been converted into the three components according to the spectral responses of the short, medium, and long wavelength retinal receptors. The conversion has been performed using empirical

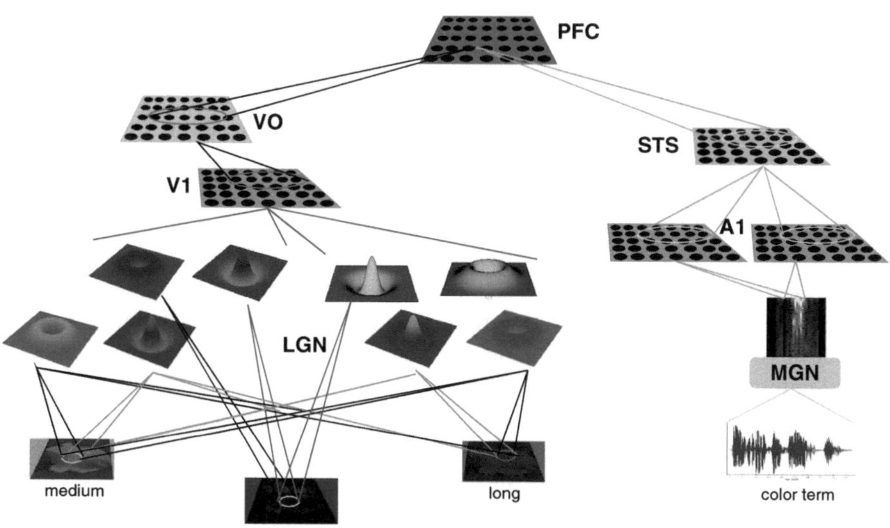


**Fig. 7.6** Scheme of the model simulating the semantics of color terms. The components are mostly the same as those in the model in Fig. 7.1, except for the subcortical color components, which here correspond to the short, medium, and long wavelength retinal receptors

data of spectral sensitivities of cone receptors gathered by Stockman and Sharpe (2000). The set of opponent receptive fields collected in the model under the label LGN is more complex than in previous models as well, accounting for all color opponent combinations in human vision (Stockman and Brainard 2010).

As in previous experiments, the model is exposed to a variety of stimuli, in different stages of its development that to some extent parallel periods of human development, from the pre-natal stage to the initial language acquisition stage. While the early developmental phase is common to all the models, in the linguistic phase three different models are developed, corresponding to the separate effects of three different environments, Berinmo, Himba, as well as the typical, more varied yet undifferentiated visual world of western cultures. The rationale behind these experiments is that we believe that the ability humans have of seeing colors is not only strongly influenced by the biological constraints acting upon the visual system of our species, by the language spoken by our linguistic community, or more precisely, by the color terms it teaches us to use, but also by our experiences with the natural world, or the "visual diet" provided to us by our natural environment. We therefore, thought it interesting to investigate the possible impact the environment might have in the cases of Berinmo and Himba, since the landscapes in which the two groups live are drastically different. In the pre-linguistic phase of development of the model, natural images are used as stimuli for the visual pathway, with three variations. A neutral one lacks dominant hues, and is typical of many urban environments in modern cultures, where the most common objects and scenes seen by newborns are man made, with a wide range of colors, prevailing over the natural hue biases of the natural environment, if any.

It has been built by taking random pictures from the Flowers and Landscape collections of the McGill Calibrated Color Image Database (Olmos and Kingdom 2004).

The other two environments are those typical of Berinmo and Himba that contrary to the neutral urban environment, are dominated by specific ranges of hues. The Berinmo environment is the luxurious vegetation of Papua New Guinea, along the large Sepik river, with villages found under the shadow of tall trees. The Himba people live in the open rocky desert lands of Northern Namibia, dominated by warm earth-colored hues.

The contrast between the three sets of visual stimuli can be appreciated by the examples given in Fig. 7.7. All pictures of the Berinmo and Himba environments are courtesy of Debi Roberson, and are shots she took during her color terms investigations of these people. The analysis of the spectral differences between the three sets is in (Plebe and De la Cruz 2014). For the linguistic development we used, as in Sect. 6.1, a large set of the most common English terms for the western-born model, and the same set of words in Spanish for the other two models. Of course there is no available synthetic speaking software for Himba and Berinmo, and it is not essential for the experiment. In this phase there is no interaction between the linguistic and the visual inputs. At the end of this stage, types of organization are found in the lower maps that enable the performance of processes that are essential to vision, with complete mapping of hues in the VO area, for all environment grown

**Fig. 7.7** Examples of the pictures used to train three different versions of the model, adapted for different environments. On the *left*, neutral pictures from the Flowers and Landscape collections, in the *middle*, the New Guinea – Sepik river environment (Courtesy of Debi Roberson), on the *right*, the Namibia – Kunene district environment (Courtesy of Debi Roberson)

models, with a slight lack of blue, and a marginal increase of red-sensitive areas for the Namibia environment. More details on the lower maps are in (Plebe and De la Cruz 2014).

The third phase of development reproduces events reminiscent of the World Color Survey protocol (Cook et al. 2005) used in testing color perception across cultures. There is a standard collection of rectangular chips, colored according to the Munsell (1912) color space, one chip at the time is shown to the subject, asking him to name the color. The model is presented with one chip at a time, but the color name in this case is heard, and the utterance corresponding to the known color term in the chosen language, associated with the Munsell color identifier of the patch. Figure 7.8 shows the regions in Munsell color space, where all the hues of the chips used in the experiments belong. For each sample hue, three variations in saturation, and three retinal poses of the colored chip are used. The developed models are then tested using our standard population analysis (see Sect. 4.3), applied to the model map PFC, the results are in Fig. 7.9. We can see that for each language, the basic colors span the entire PFC map evenly, crossing the multiple hue domains in which the color is represented. All units that are not showing any coding of single basic colors, will be activated by more than one, and therefore contribute to percepts at a finer level than basic categories. The share of PFC map coding units for basic colors is highly different between languages, even for colors that share part of the Munsell space. Finally, the same basic model was used in simulating a psychological discrimination task, to check for an effect like that of human categorical perception. The Berinmo, Himba, and English versions have been exposed to sets of triads of color stimuli, where two of the stimuli lie within one linguistic category, while the third belongs to a different category.The boundary categories are blue-green for English, nol-wor for Berinmo, and dumbu-burou for Himba. The exact values of the colors used match those used by Roberson et al. (2005). In evaluating

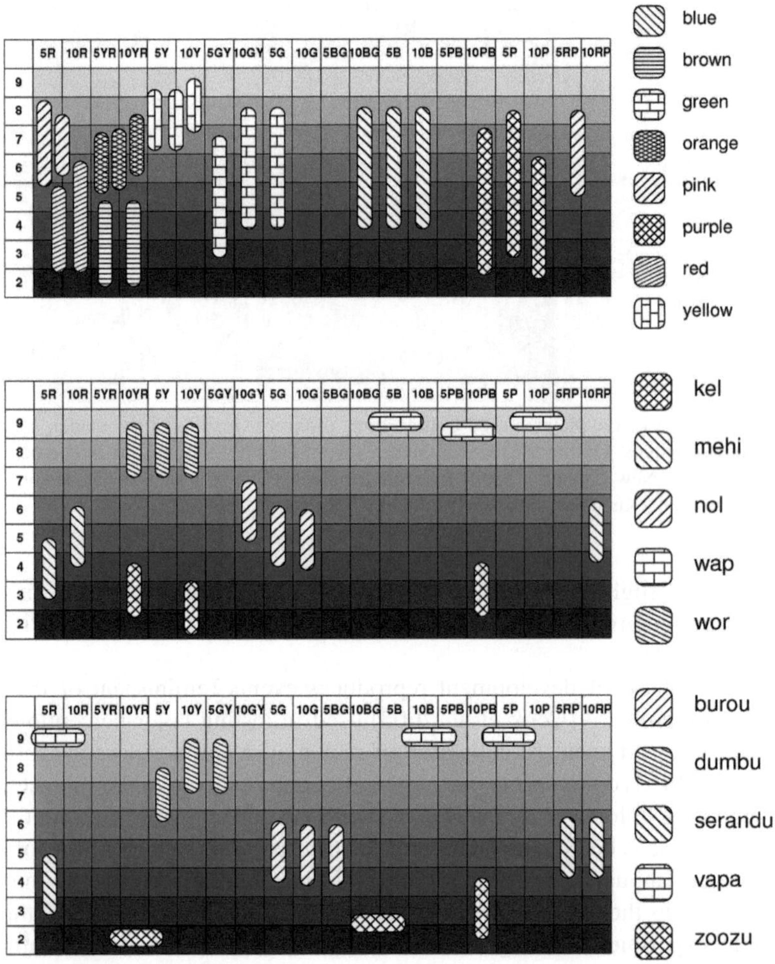


**Fig. 7.8** The basic colors of the three environments (on the left), and the localization in the Munsell charts, of all samples used in the model. On the *top* the English color terms, Berinmo in the *middle*, and Himba at the *bottom*

the triadic tests, the activities in the model PFC map are compared for similarity using all nodes, without taking into account their partition in population coding for categories. Being $w_1, w_2$ the two hues of triad within the same linguistic category, and $o$ the hue outside the boundaries of that category, the similarity judgment is computed as follows:

$$j\left(\langle w_1, w_2, o\rangle\right) = \begin{cases} 1 & \text{if } \langle w_1, w_2\rangle = \underset{\langle i\in\{w_1,w_2,o\}, j\in\{w_1,w_2,o\}, i\neq j\rangle}{\arg\min} \left\| \mathbf{x}^{\mathrm{PFC}}(i) - \mathbf{x}^{\mathrm{PFC}}(j) \right\| \\ 0 & \text{otherwise} \end{cases} \tag{7.12}$$

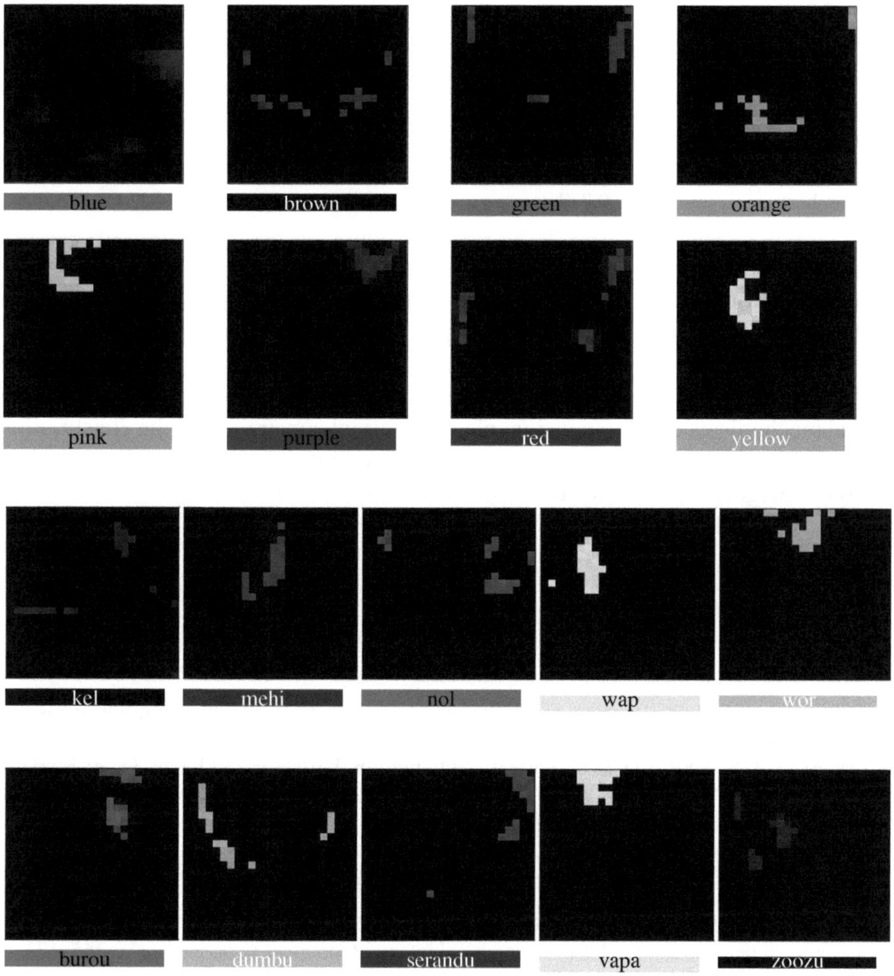


**Fig. 7.9** Population coding of the basic colors in the PFC map, for the three culture in which the models have been trained: English (*top two rows*), Berinmo (*middle*) and Himba (*bottom*)

where $\mathbf{x}^{\text{PFC}}(c)$ is the vector composed by all the activation in the PFC map, in response to the color stimulus $c$. Equation (7.12) values as 1 a response in agreement with the prediction that the two color stimuli within category are judged more similar than that crossing the category boundary, and values as 0 a different response. The average response for a single triad of colors is given by the mean value of Eq. (7.12) repeated for all combinations of samples of the same hues. The summary of the average similarity judgments for all triads and for all the three culturally-grown models, is given in Table 7.3. Results show a clear categorical perception effect for all languages, with the blue-green boundary, in

**Table 7.3** Results of the mean similarity judgments in the categorical perception tests. The three columns are the responses of the different models, grown according to the Himba, Berinmo, and English cultures

| | Himba | Berinmo | English |
|---|---|---|---|
| Blue-green | 0.443 | 0.505 | 0.781 |
| nol-wor | 0.266 | 0.790 | 0.550 |
| dumbu-burou | 0.777 | 0.604 | 0.532 |

particular, strongly affecting the English model, while having an indifferent effect on Himba and Berinmo.

Summing up, the neurosemantics of color terms simulated by the models here discussed, seem to corroborate the relativist view. We certainly believe that the neural mechanisms of human vision, place important constraints on the construction of a lexical system of color terms, but these constraints would allow a large variety of color categories, which would depend on the history of our languages as well as our cultures, and perhaps partly, on our physical environments.

## References

Aboitiz, F., Garcia, R. R., Bosman, C., & Brunetti, E. (2006). Cortical memory mechanisms and language origins. *Brain and Language, 98*, 40–56.

Atzori, M., Lei, S., Evans, D. I. P., Kanold, P. O., Phillips-Tansey, E., McIntyre, O., & McBain, C. J. (2001). Differential synaptic processing separates stationary from transient inputs to the auditory cortex. *Neural Networks, 4*, 1230–1237.

Baddeley, A. (1992). Working memory. *Science, 255*, 556–559.

Berlin, B., Kay, P. (1969). *Basic color terms. Their universality and evolution.* Berkeley: California University Press.

Clifford, A., Franklin, A., Davies, I. R., & Holmes, A. (2009). Electrophysiological markers of categorical perception of color in 7-month old infants. *Biological Cybernetics, 71*, 165–172.

Cook, R. S., Kay, P., & Regier, T. (2005). The world color survey database: History and use. In H. Cohen & C. Lefebvre (Eds.), *Handbook of categorization in cognitive science* (pp. 223–242). Amsterdam: Elsevier.

Darwin, C. (1877). A biographical sketch of a young infant. *Kosmos, 1*, 367–376.

Franklin, A., Drivonikou, V., Bevis, L., Davies, I., Kay, P., & Regier, T. (2008). Categorical perception of color is lateralized to the right hemisphere in infants, but to the left hemisphere in adults. *Proceedings of the Natural Academy of Science USA, 105*, 3221–3225.

Franklin, A., Wright, O., & Davies, I. (2009) What can we learn from toddlers about categorical perception of color? Comments on Goldstein, Davidoff, and Roberson. *Journal of Experimental Child Psychology, 102*, 239–245.

Fuster, J. M. (2001). The prefrontal cortex–an update: Time is of the essence. *Neuron, 30*, 319–333.

Gasser, M., & Smith, L. B. (1998). Learning nouns and adjectives: A connectionist account. *Language and Cognitive Processes, 13*, 269–306.

Gentner, D. (1978). On relational meaning: The acquisition of verb meaning. *Cognitive Development, 49*, 988–998.

Heider, E. R., & Olivier, D. C. (1972). The structure of the color space in naming and memory for two languages. *Cognitive Psychology, 3*, 337–354.

Kay, P., Berlin, B., & Merrifield, W. (1991). Biocultural implications of systems of color naming. *Journal of Linguistic Anthropology, 1*, 12–25.

Komarova, N. L., & Jameson, K. A. (2008). Population heterogeneity and color stimulus heterogeneity in agent-based color categorization. *Journal of Theoretical Biology, 253*, 680–700.

Lucy, J. A. (1997). The linguistics of color. In C. L. Hardin & Maffi, L. (Eds.), *Color categories in thought and language* (pp. 320–346). Cambridge: Cambridge University Press.

Mintz, T. H., & Gleitman, L. R. (2002). Adjectives really do modify nouns: The incremental and restricted nature of early adjective acquisition. *Cognition, 84*, 267–293.

Munsell, A. H. (1912). A pigment color system and notation. *The American Journal of Psychology, 23*, 236–244.

Nayar, S., & Murase, H. (1995). Visual learning and recognition of 3-d object by appearence. *International Journal of Computer Vision, 14*, 5–24.

Olmos, A., & Kingdom, F. A. (2004). McGill calibrated colour image database. http://tabby.vision.mcgill.ca.

Özgen, E., & Davies, I. R. L. (2002). Acquisition of categorical color perception: A perceptual learning approach to the linguistic relativity hypothesis. *Journal of Experimental Psychology: General, 131*, 477–493.

Plebe, A., & De la Cruz, V. M. (2014). Color seeing and speaking – effects of biology, environment and language. In W. Anderson, Biggam, C. P., Hough, C., & Kay, C. (Eds.), *Colour studies. A broad spectrum* (pp. 291–306). Amsterdam: John Benjamins.

Plebe, A., Mazzone, M., & De La Cruz, V. M. (2011). Colors and color adjectives in the cortex. In C. Biggam, C. Hough, C. J. Kay, & D. Simmons (Eds.), *New directions in colour studies* (pp. 415–428). Amsterdam: John Benjamins.

Plebe, A., De la Cruz, V. M., & Mazzone, M. (2013). In learning nouns and adjectives remembering matters: A cortical model. In A. Villavincencio, T. Poibeau, A. Korhonen, & A. Alishahi (Eds.), *Cognitive aspects of computational language acquisition* (pp. 105–129). Berlin: Springer.

Regier, T., Kay, P., & Khetarpal, N. (2007). Color naming reflects optimal partitions of color space. *Proceedings of the Natural Academy of Science USA, 104*, 1436–1441.

Roberson, D., & Hanley, J. R. (2010). Relatively speaking – An account of the relationship between language and thought in the color domain. In B. C. Malt & P. Wolff (Eds.), *Words and the mind* (pp. 183–198). Oxford: Oxford University Press.

Roberson, D., Davidoff, J., & Davies, I. R. (2000). Colour categories are not universal: Replications and new evidence from a stone-age culture. *Journal of Experimental Psychology: General, 129*, 369–398.

Roberson, D., Davidoff, J., Davies, I. R., & Shapiro, L. R. (2004). The development of color categories in two languages: A longitudinal study. *Journal of Experimental Psychology: General, 133*, 554–571.

Roberson, D., Davidoff, J., Davies, I. R., & Shapiro, L. R. (2005). Color categories: Evidence for the cultural relativity hypothesis. *Cognitive Psychology, 50*, 378–411.

Sandhofer, C. M., & Smith, L. B. (2001). Why children learn color and size words so differently: Evidence from adults' learning of artificial terms. *Journal of Experimental Psychology, 130*, 600–620.

Saunders, B., & Van Brakel, J. (1997). Are there non-trivial constraints on colour categorisation? *Behavioral and Brain Science, 20*, 167–232.

Steels, L., & Belpaeme, T. (2005). Coordinating perceptually grounded categories through language: A case study for colour. *Behavioral and Brain Science, 28*, 469–529.

Stockman, A., & Brainard, D. H. (2010). Color vision mechanisms. In M. Bass (Ed.), *OSA handbook of optics* (pp. 11.1–11.104). New York: McGraw Hill.

Stockman, A., & Sharpe, L. T. (2000). The spectral sensitivity of the middle- and long-wavelength-sensitive cones derived from measurements in observers of known genotype. *Vision Research, 40*, 1711–1737.

Vallar, G., & Shallice, T. (Eds.) (2007) *Neuropsychological impairments of short-term memory*. Cambridge: Cambridge University Press.

Yendrikhovskij, S. N. (2001). Computing color categories from statistics of natural images. *The Journal of Imaging Science and Technology, 45*, 409–417.

Zhou, K., Mo, L., Kay, P., Kwok, V. P. Y., Ip, T. N. M., Tan, L. H. (2010). Newly trained lexical categories produce lateralized categorical perception of color. *Proceedings of the Natural Academy of Science USA, 107*, 9974–9978.

# Chapter 8
# Toward a Neurosemantics of Moral Terms

**Abstract** In all the aspects of semantics explored in the previous chapters, neural computation is today a fairly well established approach, with the models discussed not being the only ones. On the other hand, a neurocomputational approach has not been yet established for terms related to morality. The main reason is that empirical brain information on moral processing is still in its early infancy. As has been found with other aspects of word meaning, neuroscientific investigations such as those undertaken by Greene et al. (2001), have shown that there is a relatively consistent set of brain areas that are engaged during moral reasoning, as well as during emotional reactions and decision making in general (Casebeer and Churchland 2003; Moll et al. 2005).

The models presented in this chapter can, therefore, be considered as pioneering works, or first steps in the enterprise of the neurosemantics of morality. The first model lacks linguistic capabilities and is a collection of a series of brain areas that learn the emotional component that contributes to the evaluation of potential actions, and then makes decisions. The second model combines the morally relevant areas of the first, with the auditory pathway that processes linguistic utterances, and simulates the emergence of the meaning of moral terms.

## 8.1 Ethics, Semantics, and Computation

Not only does the study of human morality appear distant from computational issues, it comes out looking as if not a semantic problem at all. It is not just a matter of common sense, there are outstanding philosophical positions that warn that morality resists formalization within semantic theories. The famous "open-question argument" put forward by Moore (1903) allegedly dismissed any proposition that inferred moral value from natural properties, as a logical fallacy. Even if his argument was basically flawed (Frankena 1939), Moore's non-naturalism profoundly shaped moral philosophy in his century. Even more compelling is an older similar claim made by Hume (1740), that no "ought" can be derived by an "is". Scientific facts are descriptive, while moral facts are prescriptive, and since it is impossible to deduce a statement that has obligatory force from statements that are purely descriptive, moral theories cannot be pursued within semantic theories.

 
A. Plebe, V.M. De La Cruz, *Neurosemantics*,
Studies in Brain and Mind 10, DOI 10.1007/978-3-319-28552-8_8

The first attempts to break with such positions came with the analytic tradition. For Hare (1952, p. III) "Ethics, as I conceive it, is the logical study of the language of morals", and he was one of the first to engage in the project of including morality within formal logic. In his view, moral sentences share similarities with imperatives, with the difference of being more universal. Both belong to the general class of prescriptive languages, for which meaning comes in two components: the *phrastic* which captures the state to be the case, or command to be made the case, and the *neustic* part, that determines the way the sentence is put together by the speaker. While Hare did not provide technical details of his idea for prescriptive languages, in the same years Von Wright (1951) developed deontic logic, the logical study of normative concepts in language, with the introduction of the monadic operators $O(\cdot)$, $F(\cdot)$, and $P(\cdot)$ for expressing obligation, prohibition and permission. It is well known that the many attempts in this direction engender a set of logical and semantic problems, with the most severe being the Frege-Geach (1965) embedding problem. Since the semantics of moral sentences is determined by a non-truth-apt component, like Hare's neustic, it is unclear how they can be embedded into more complex propositions, such as conditionals for example.

### 8.1.1 *Logic, Morality and the Mind*

Not surprisingly, the line of research on morality within logic was detached from real mental processes required for moral cognition, as has been discussed in Sect. 5.1, but details of this research are not of interest here. An interesting exception can be found among the contemporary proponents of expressivism, the theory that moral judgments express attitudes of approval or disapproval. Since this sort of attitude pertains to the mental world, and are driven by emotional motivations, attempts to provide a more precise account of the meaning of moral sentences should require a step further, towards what Wedgwood (2007) calls a psychologistic semantics.

A quick glance will be given at the two best available attempts. Blackburn (1988) introduced variants of the deontic operators, like $H!(\cdot)$ and $B!(\cdot)$, that merely express attitudes regarding their argument: "Hooray!" or "Boo!". Every expressive operator has its descriptive equivalent, given formally by the $|\cdot|$ operation, for example

$$H!\,(|B!(X)| \rightarrow |B!(\text{teach to}X)|) \tag{8.1}$$

conveys the approbation that action $X$ deserves disapprobation, and therefore it is disapproved to teach it. This step is a tentative response to the Frege-Geach problem.

Gibbard (1990) frames his proposal in possible worlds semantics rather than deontic logic, and defines an equivalent expressivist friendly concept, that of *factual-normative world*:

$$\langle W, N \rangle \tag{8.2}$$

where $W$ is an ordinary Kripke-Stalnaker possible world, while $N$, the system of norms, is characterized by a family of predicates like $N$-forbidden, $N$-required. If a moral sentence $S$ is $N$-permitted in $\langle W, N \rangle$ then it is said to hold in that factual-normative world.

Both proponents acknowledge the need of moving towards a mental inquiry, for Blackburn (1998, p. 59) "Expressivism requires a naturalistic story of the state of mind of valuing something." Gibbard (1990) is well aware that "talk of a set of factual-normative worlds seems psychologically farfetched. How could that be what anyone has in mind when he thinks normatively?" (p. 97). "These are versions of wider questions of what logic has to do with meaning, and what meaning has to do with things that go on in the mind." (p. 100). However, their aim never did translate into an effective attempt to embed genuine mental processes in a logic system.

### 8.1.2 The Linguistic Analogy

A recent computational approach to morality is worth mentioning, one that put high claims on being, unlike deontic logic, a faithful reconstruction, even if approximated, of mental mechanisms. It has been proposed by Mikhail (2000), and is based on a Chomskian-like grammar (see Sect. 5.2.1). This modeling effort draws attention to what is called the "linguistic analogy", and become known as the *Universal Moral Grammar* (UMG). It is focused on a single moral dilemma, the famous "trolley" (Foot 1967; Greene et al. 2001), in which a decision has to be made to save people in danger of death, with the possible side effect of killing others. It implements a case of the so-called doctrine of the double effect, which differentiates between harm caused as means and harm caused as a side effect (Thompson 1985). Mikhail takes as fixed the main structure and circumstances of the standard trolley paradigm, and contrived several subtle variations, in order to extend the two classical cases to twelve subcases (Mikhail 2009). He submitted all the variations of the dilemmas to subjects in a series of experiments, collecting their responses with terms like "permissible", "forbidden", "obligatory". The model he developed has the purpose of giving the same response as that most commonly given by the subjects. Much of his model is based on an old theory, again inspired by generative grammar, that of Goldman (1970) of describing human actions by a sort of syntactic tree. A complex act is "generated" by more basic act constituents according to a set of rewriting-like rules. This formatted structural description of the situation and of the potential action is the input to a sort of grammatical parser (Dowty 1985), which takes as input a formal grammar and a sentence, and gives as output the decision about the grammaticality of the sentence. In this case the output is the decision if the potential action is permissible, forbidden, or obligatory.

Unlike the case of Universal Grammar in linguistics, where hundreds of scholars have contributed equally to developments of the theory as applied to specific aspects of language, and to theoretical discussions, the situation inside UMG is quite peculiar. There have been no other researchers who have attempted to further develop Mikhail's approach, but many have theorized about it, in particular Hauser (2006a,b), but see also Harman (1999), Dwyer (2008), and Roedder and Harman (2010). The idea that morality works as a universal grammar has met with several critics as well, such as Nichols (2005), Dupoux and Jacob (2007b), Mallon (2008), Prinz (2008), and Sterelny (2010). There are reasons for sympathizing with the model proposed by Mikhail, first of all, for being a computational account of how morality works, second, it could be a possible competitor with classic and deontic logic in the modeling of legal rules (Holton 2011). Unfortunately, this is not his aim, and even less is it Hauser's or of any of the other proposers of UMG, who aspire to use it as a description of the mental processes in human morality: "The moral grammar hypothesis holds that ordinary individuals are intuitive lawyers, who possess tacit or unconscious knowledge of a rich variety of legal rules, concepts, and principles, along with a natural readiness to compute mental representations of human acts and omissions in legally cognizable terms" (Mikhail 2009, p. 29). There are several reasons that make the idea of morality like a syntactic grammar untenable. A first order of arguments are the weaknesses of the analogy between morality and language in itself (Dupoux and Jacob 2007a,b). A second, possibly even more serious, is that moral grammar suffers from the same drawback that has affected generative linguistics since its beginnings, the contradiction concerning the abstract mathematical nature of the entities inside grammatical theory, and their assumed psychological valence. It is the failure of generative grammar as a cognitive theory, that prompted cognitive semantics, discussed in Sect. 5.2. Unfortunately, this serious problem seems to have being ignored by UMG proponents. Even conceding to defenders of UMG a validity of the linguistic analogy, their choice among linguistic theories has been that of generative grammar, which has failed to match with cognition. In addition, the modeling approach taken by UMG is sorely missing some of the most important aspects of morality recent empirical studies have brought to attention, such as the emotional component, which will be addressed in Sect. 8.2.

Needless to say, Mikhail's computational model fails entirely in passing the model-mechanism-mapping criteria for an explanation of brain functioning in morality.

### *8.1.3 Neurocomputational Pieces*

In trying to embed the domain of moral terms within our unified neurosemantic conception, we are not starting from scratch. Even if neurocomputational approaches to morality are still lacking, there are a number of pieces that exist, whose purpose is that of modeling brain functions, which are in close relation to morality. The most relevant are the neurocomputational models of decisions and emotions.

For both aspects, an important framework of reference is the same found when argumenting our main principle of coincidence detection: reinforcement learning (see Sect. 3.2.1). Solutions to theoretical reinforcement learning using neuronlike elements were first proposed by Barto and Sutton (1982) and Barto et al. (1983), and have been gradually fitted into the biology of neuromodulation and forebrain circuits implicated in decision making (Daw et al. 2002; Doya 2002; Dayan 2008; Bullock et al. 2009). The brain continuously faces making decisions in everyday life, from simple motor control up to long term planning, and few of them specifically involve moral judgments, but a small class of reinforcement learning models verge on components of moral behavior. The GAGE model proposed by Wagar and Thagard (2004), named with reference to the historical case of Phineas Gage, assembles groups of artificial neurons corresponding to the ventromedial prefrontal cortex, the hippocampus, the amygdala, and the nucleus accumbens. It hinges on the somatic-marker idea of Damasio (1994), feelings that have become associated through experience with the predicted long-term outcomes of certain responses to a given situation. GAGE was tested in a simplified version of the Iowa Gambling Task (Bechara et al. 1994), selecting cards from two decks. One can give larger immediate rewards, but a long term overall loss, the other gives smaller rewards but a gain in the long term. The model was able to learn to decide for the long term reward, by virtue of the associations made between the ventromedial prefrontal cortex and the amygdala, of the experienced loss.

GAGE implementation of somatic-markers was based on Hebbian learning only, without reinforcement learning, which was adopted instead in ANDREA (Litt et al. 2006, 2008), a model where the orbitofrontal cortex, the dorsolateral prefrontal cortex and the anterior cingulate cortex, interact with basal ganglia and the amygdala. This model was designed to reproduce a well known phenomenon in economics: the common hypersensitivity to losses over equivalent gains, analyzed in the prospect theory of Kahneman and Tversky (1979). The asymmetric evaluation of gains and losses is simulated in ANDREA at the output of the orbitofrontal cortex, under the effect of the amygdala, conveying emotional arousal. Thagard and Aubie (2008) announced EMOCON, a sophisticated model that incorporates ideas from ANDREA and GAGE, along with simulations of sensorial inputs that were lacking in both previous models. One challenging target of this future model, is the simulation of emotional consciousness.

The overall architecture of the models of Thagard and his group has several similarities with those of Frank and Claus (2006) and Frank et al. (2007), in which the orbitofrontal cortex interacts with the basal ganglia, but more oriented to dichotomic on/off decisions.

## 8.2 The Emotional Coding of Norms

Before coming up with a neurocomputational model of morality, one is faced with the problem of establishing a working definition of what morality is. Coming up with a precise definition of morality is exactly one of the main endeavors of moral philosophy. Establishing a clear cut division between moral decisions and everyday social problem-solving in nonmoral decisions, or between moral norms and social conventions, is neither a simple nor straightforward task. The road undertaken in our neurosemantics model of morality, in addition to being based as much as possible on relevant brain facts, is also necessarily rooted in a number of theoretical positions that will be spelled out now.

### 8.2.1 *Moral Behavior Is Learned Emotion*

First, at the core of the model are the emotion brain centers involved in values and decisions, because we embrace the idea that moral cognition is emotional in nature. It is a view, that is part of a philosophical tradition that goes back to Hume (1740). Among its most authoritative contemporary defenders we have Prinz (2006a, 2008) and Nichols (2004), whose detailed analyses dispenses us from discussing the full set of supporting motivations. The emotional basis of morality has been ascertained in a large number of neuroscientific studies. Moll et al. (2005) reported the remarkable correlation between damage causing deficient emotional engagement, and impairments in moral judgments, as in the case of the ventromedial prefrontal cortex. Dysfunction in these areas is also typical in psychopathy, characterized by poor moral behavior, together with dysfunction in the amygdala (Blair 2007) and in the orbitofrontal cortex (Blair 2010). Moral judgments and emotions seem to coincide in the brain, with structures in addition to those just mentioned, including the insula, anterior cingulate cortex, the temporal pole, and the medial frontal gyrus (Moll et al. 2008). Additional evidence comes from studies showing that manipulating emotions can influence moral judgments (Schnall et al. 2008). Cameron et al. (2013) demonstrate that it is even possible, with specific training, to make fine-grained distinctions between emotions that are incidental to the actions being judged versus emotions that are integral to them, discounting inappropriate emotions while making moral judgments.

The second fundamental essence of morality is that it is learned emotion. Even if emotion is grounded in the same neural equipment evolved in humans for sociality and cognition, evolutionary theory falls short of explaining any of our specific moral values. Current neurobiological evidence does not indicate that something like a set of moral rules exists in the brain, but there is certainly a set of strong biological biases towards sociality and the care taking of others that we associate with morality, but they are too general and unconstrained to guide the variety of specific behaviors that are involved and displayed in human morality (Casebeer and Churchland 2003; Moll et al. 2008; Suhler and Churchland 2011; Churchland 2011; Young and Dungan 2012). In addition, all the areas involved in morality are highly plastic. The frontal structures belong to the most critical region for learning, storing, and binding social knowledge to contextual elements. In a study with participants aged between 4 and 37 years Decety et al. (2012) found an age-related increase in activity in the ventromedial prefrontal cortex, as well as increased functional connectivity between this region and the amygdala, in response to dynamic visual stimuli depicting moral transgressions.

### 8.2.2 Morality Is Not a Single Mechanism

What comes under the label of morality, when examined closely, looks much more like a collection of different patterns of behavior, rather than a monolithic set of beliefs. It comes as a natural consequence of the essence of morals as emotions. There are distinct basic emotions, in which several classes of moral situations find their place. Stich (2006) has cogently proposed the idea of how dissociated the cognitive pieces that make up morality are, interpreting morality as being more like a kludge than an elegant machine.

One of the first taxonomies attempted on an empirical basis, was the psychological study of Rozin et al. (1999), in which subjects were presented with vignettes that depicted either a clear harm, an instance of disrespect, or a case of something we tend to regard as polluting the body. Subjects were asked to identify the appropriate emotional response. Rozin et al. proposed a model to explain the results, called CAD for the three emotions: contempt, anger, and disgust, but also for the three related classes of moral codes: community, autonomy, and divinity. The CAD model is an important achievement, still very influential in moral cognition, despite the fact that the exact definition of the classes of emotions and the related moral codes have been debated. Prinz (2008, pp. 73–75) proposes a first main classification in other- and self-directed moral emotions, with only two basic emotions in the first: anger and disgust. Rozin's contempt is a blending of those two. The reflexive basic moral emotions are guilt and shame.

Recent neurocognitive experiments have confirmed that morality is not a wholly unified faculty, but rather, instantiated in partially dissociable neural systems, that are engaged differentially depending on the kind of emotion elicited by the moral transgression (Parkinson et al. 2011). Guilt, which is likely to follow when inflicting physical harm or taking something from a member of the group, is the type of emotion on which the models here described are based. The context, in which the moral violation takes place is the attempt by the model to steal someone's food. This action will provoke the angry facial expressions of the victim, seen by the model, and to which it will react with an emotion of guilt.

### 8.2.3 The Moral Neural Engine

The overall architecture of the first model aimed at simulating the acquisition of a single moral norm on emotional basis, is shown in Fig. 8.1. It is composed by a minimum set of maps equivalent, in a highly simplified form, to the brain areas making up the "moral neural engine", limited to a violation inducing a guilt emotion. A word of caution concerning the adherence of the components of the model, with respect to the brain areas with the same lables, already mentioned in previous experiments, has to be repeated with emphasis here. Areas like OFC, Amyg, vmPFC, VS are engaged in the brain in a wide range of processes, that are

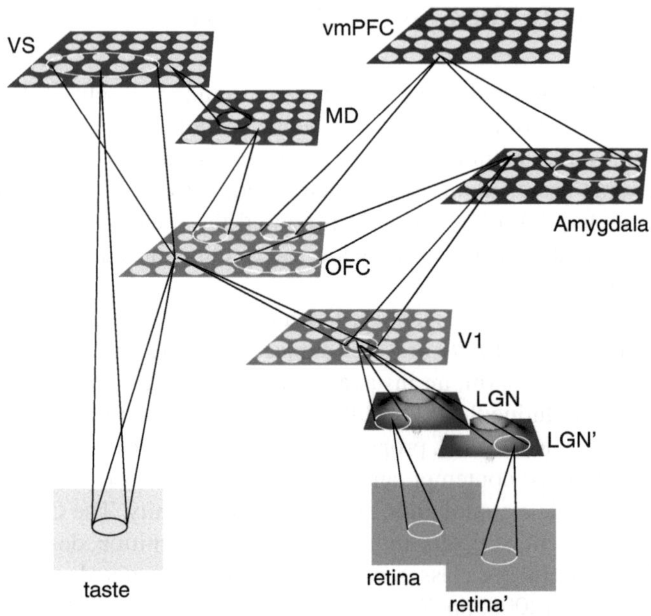


**Fig. 8.1** Overall scheme of the moral emotional model, composed by LGN (*Lateral Geniculate Nucleus*), V1 (*Primary Visual Area*), OFC (*OrbitoFrontal Cortex*), VS (*Ventral Striatum*), MD (*Medial Dorsal nucleus of the thalamus*), Amyg (*Amygdala*), vmPFC (*ventromedial PreFrontal Cortex*). LGN and Retina are replicated in LGN' and Retina' as a delayed visual scene, which may contain the angry face

ignored in this model. All that is preserved in the model is the hierarchy and the reciprocal connectivity, together with the limited functions performed in the model which, as will be discussed next, have been demonstrated in the brain.

There are two main circuits that learn the emotional component that contributes to the evaluation of potential actions. The first, comprises the orbitofrontal cortex, with its processing of sensorial information, reinforced with positive perspective values by the loop with the ventral striatum and the dopaminergic neurons. The second, shares the representations of values from the orbitofrontal cortex, which are evaluated by the ventromedial prefrontal cortex against conflicting negative values, encoded by the closed loop with the amygdala. Now all the components will be described, with a justification of their role in morality, from neurophysiological evidence and their mathematical representation in the model. In the equations, for the sake of readability, the following symbols will be used for the subcortical signals:

- $\odot$ the output of the LGN at the time when seeing the main scene;
- $\circledcirc$ the output of the LGN deferred in time, when a possibly angry face will appear;
- $\boxdot$ the taste signals;
- $\circledast$ the output of the medial dorsal nucleus of the thalamus.

A key map in the model is the orbitofrontal cortex, which in the brain receives a varied assortment of sensorial information from the visual stream, taste, olfactory, auditory, and somatosensory inputs (Rolls 2004). There are neurons in the orbitofrontal cortex that respond differently to visual objects depending on their reward association, and one of the primary reinforcements is taste (Rolls et al. 1996). In addition, there is a population of orbitofrontal neurons which respond to faces (Rolls et al. 2006), and some respond specifically to facial expressions. The crucial role played by the orbitofrontal cortex in social decision making was discovered through the observation of patients with lesions (Damasio 1994; Bechara et al. 1994). Its specific relevance for morality is controversial. According to Greene and Haidt (2002) this area might perform a general regulative function, in which affective information guides approach and avoidance behavior in both social and non-social contexts. However, the orbitofrontal cortex is almost always involved in moral cognition (Moll et al. 2005). For Prehn and Heekeren (2009) the role of the orbitofrontal cortex in moral judgment is the representation of the expected value of possible outcomes of a behavior in regards to rewards and punishments. The equation of the activation of a unit in the OFC layer of the model is the following:

$$
\begin{aligned}
x^{(\mathrm{OFC})} = f\Big( & \gamma_{\mathrm{A}}^{(\mathrm{OFC}\leftarrow \mathrm{V1})} \mathbf{a}_{r_{\mathrm{A}}}^{(\mathrm{OFC}\leftarrow \mathrm{V1})} \cdot \mathbf{v}_{r_{\mathrm{A}}}^{(\mathrm{V1})} + \gamma_{\mathrm{A}}^{(\mathrm{OFC}\leftarrow \circledcirc)} \mathbf{a}_{r_{\mathrm{A}}}^{(\mathrm{OFC}\leftarrow \circledcirc)} \cdot \mathbf{v}_{r_{\mathrm{A}}}^{(\circledcirc)} \\
& + \gamma_{\mathrm{A}}^{(\mathrm{OFC}\leftarrow \boxdot)} \mathbf{a}_{r_{\mathrm{A}}}^{(\mathrm{OFC}\leftarrow \boxdot)} \cdot \mathbf{v}_{r_{\mathrm{A}}}^{(\boxdot)} + \gamma_{\mathrm{B}}^{(\mathrm{OFC}\leftarrow \circledast)} \mathbf{b}_{r_{\mathrm{B}}}^{(\mathrm{OFC})} \cdot \mathbf{v}_{r_{\mathrm{B}}}^{(\circledast)} \\
& + \gamma_{\mathrm{E}}^{(\mathrm{OFC})} \mathbf{e}_{r_{\mathrm{E}}}^{(\mathrm{OFC})} \cdot \mathbf{x}_{r_{\mathrm{E}}}^{(\mathrm{OFC})} - \gamma_{\mathrm{H}}^{(\mathrm{OFC})} \mathbf{h}_{r_{\mathrm{H}}}^{(\mathrm{OFC})} \cdot \mathbf{x}_{r_{\mathrm{H}}}^{(\mathrm{OFC})} \Big)
\end{aligned}
\tag{8.3}
$$

There are three sensorial afferents: $\mathbf{v}_{r_{\mathrm{A}}}^{(\mathrm{V1})}$ from the visual cortex V1, $\mathbf{v}_{r_{\mathrm{A}}}^{(\odot)}$ from the lateral geniculate nucleus of the thalamus, and the taste sensorial input $\mathbf{v}_{r_{\mathrm{A}}}^{(\boxdot)}$, each in a sensorial area $r_{\mathrm{A}}$ corresponding to the receptive field of the unit in OFC. A fourth afferent, $\mathbf{v}_{r_{\mathrm{B}}}^{(\circledast)}$, is the diffuse projection from MD, carrying dopamine signaling from the loop that will be described next, its equation will be given in (8.6).

In this model the visual pathway is not as detailed as in Sect. 6.1, where the semantics were closely related with visual features. In this case it is simplified in a single area, V1, with the following equation:

$$
x^{(\mathrm{V1})} = h\left( \gamma_{\mathrm{A}}^{(\mathrm{V1}\leftarrow \odot)} \mathbf{a}_{r_{\mathrm{A}}}^{(\mathrm{V1}\leftarrow \odot)} \cdot \mathbf{v}_{r_{\mathrm{A}}}^{(\odot)} + \gamma_{\mathrm{E}}^{(\mathrm{V1})} \mathbf{e}_{r_{\mathrm{E}}}^{(\mathrm{V1})} \cdot \mathbf{x}_{r_{\mathrm{E}}}^{(\mathrm{V1})} - \gamma_{\mathrm{H}}^{(\mathrm{V1})} \mathbf{h}_{r_{\mathrm{H}}}^{(\mathrm{V1})} \cdot \mathbf{x}_{r_{\mathrm{H}}}^{(\mathrm{V1})} \right)
\tag{8.4}
$$

The output of LGN is the same given by Eqs. (6.1) and (6.2). Note that the visual input Retina, and the LGN content, are duplicated, as a way to process two visual scenes shifted in time, it is necessary in the case when the action of the model is causing the appearance of the angry face, as will be explained in Sect. 8.2.4.

The Ventral Striatum, VS, in the brain includes the nucleus accumbens and the broad continuity in the basal ganglia between the caudate nucleus and putamen, and plays a major role in various aspects of reward processes and motivation. Cortico-striatal terminals have a topographic organization, with distinct terminal fields from the orbitofrontal cortex, the ventromedial prefrontal cortex, and the anterior cingulate cortex. VS has a direct and reciprocal connection with the dopaminergic neurons located in the substantia nigra pars compacta and the ventral segmental area, which project back to MD, the medial dorsal nucleus of the thalamus, which in turn, close the reward circuit by projecting to the prefrontal cortex (Haber 2011). This circuit is implemented in the model by the following two equations:

$$x^{(\mathrm{VS})} = f\Bigg(\gamma_{\mathrm{A}}^{(\mathrm{VS}\leftarrow\mathrm{OFC})}\mathbf{a}_{r_{\mathrm{A}}}^{(\mathrm{VS}\leftarrow\mathrm{OFC})}\cdot\mathbf{v}_{r_{\mathrm{A}}}^{(\mathrm{OFC})} + \gamma_{\mathrm{A}}^{(\mathrm{VS}\leftarrow\boxdot)}\mathbf{a}_{r_{\mathrm{A}}}^{(\mathrm{VS}\leftarrow\boxdot)}\cdot\mathbf{v}_{r_{\mathrm{A}}}^{(\boxdot)} + \gamma_{\mathrm{E}}^{(\mathrm{VS})}\mathbf{e}_{r_{\mathrm{E}}}^{(\mathrm{VS})}\cdot\mathbf{x}_{r_{\mathrm{E}}}^{(\mathrm{VS})} - \gamma_{\mathrm{H}}^{(\mathrm{VS})}\mathbf{h}_{r_{\mathrm{H}}}^{(\mathrm{VS})}\cdot\mathbf{x}_{r_{\mathrm{H}}}^{(\mathrm{VS})}\Bigg) \tag{8.5}$$

$$x^{(\circledast)} = f\left(\gamma_{\mathrm{A}}^{(\circledast\leftarrow\mathrm{VS})}\mathbf{a}_{r_{\mathrm{A}}}^{(\circledast\leftarrow\mathrm{VS})}\cdot\mathbf{v}_{r_{\mathrm{A}}}^{(\mathrm{VS})}\right) \tag{8.6}$$

The afferent signals $\mathbf{v}^{(\mathrm{OFC})}$ come from Eq. (8.3), $\mathbf{v}^{(\boxdot)}$ is the taste signal. The output $x^{(\circledast)}$ computed in (8.6) will close the loop into the prefrontal cortex with Eq. (8.3). In this equation there is a parameter, $\gamma_{\mathrm{B}}^{(\mathrm{OFC}\leftarrow\circledast)}$, which will be used in a special way during the experiments. It is a global modulatory factor of the amount of dopamine signaling for gustatory reward, and therefore, is the most suitable parameter for simulating hunger states.

The top map in the model corresponds to the ventromedial prefrontal cortex, vmPFC, this region has been shown to play a crucial role in emotion regulation and social decision making (Bechara et al. 1994; Damasio 1994). According to Hernandez et al. (2009) vmPFC stands out as the heart of neural machinery involved in emotional intelligence, the ability to reliably regulate and utilize emotional information in evaluating choices. The vmPFC has been proposed as encoding a kind of common currency enabling consistent value based choices between actions and goods of various types (Gläscher et al. 2009; Boorman and Noonan 2011). In this model vmPFC is implemented by the following equations:

$$x^{(\mathrm{vFC})} = f\Bigg(\gamma_{\mathrm{A}}^{(\mathrm{vFC}\leftarrow\mathrm{OFC})}\mathbf{a}_{r_{\mathrm{A}}}^{(\mathrm{vFC}\leftarrow\mathrm{OFC})}\cdot\mathbf{v}_{r_{\mathrm{A}}}^{(\mathrm{OFC})} + \gamma_{\mathrm{A}}^{(\mathrm{vFC}\leftarrow\mathrm{Amy})}\mathbf{a}_{r_{\mathrm{A}}}^{(\mathrm{vFC}\leftarrow\mathrm{Amy})}\cdot\mathbf{v}_{r_{\mathrm{A}}}^{(\mathrm{Amy})} + \gamma_{\mathrm{E}}^{(\mathrm{vFC})}\mathbf{e}_{r_{\mathrm{E}}}^{(\mathrm{vFC})}\cdot\mathbf{x}_{r_{\mathrm{E}}}^{(\mathrm{vFC})} - \gamma_{\mathrm{H}}^{(\mathrm{vFC})}\mathbf{h}_{r_{\mathrm{H}}}^{(\mathrm{vFC})}\cdot\mathbf{x}_{r_{\mathrm{H}}}^{(\mathrm{vFC})}\Bigg) \tag{8.7}$$

The afferent signals $\mathbf{v}^{(\mathrm{OFC})}$ come from Eq. (8.3), while $\mathbf{v}^{(\mathrm{Amy})}$, is the connection from Amygdala, the core of the negative emotional reaction in this model. It will convey to vmPFC the coded negative effect of having seen the angry face in a given context, a role corresponding to well documented brain evidence. Blair (2007) alleges that learning the basics of care-based morality relies on the crucial role of the amygdala in stimulus-reinforcement learning, and in turn this learning enables representations of conditioned stimuli within vmPFC to be linked to emotional responses. The involvement of the amygdala in the recognition of facial expressions is also well established, with different kinds of expressions clustered in different subregions, and with the strongest activation in response to direct-gazing angry faces (Boll et al. 2011).

The activation of units in the artificial amygdala component is given by:

$$x^{(\mathrm{Amy})} = f\Bigg(\gamma_{\mathrm{A}}^{(\mathrm{Amy}\leftarrow\mathrm{OFC})}\mathbf{a}_{r_{\mathrm{A}}}^{(\mathrm{Amy}\leftarrow\mathrm{OFC})}\cdot\mathbf{v}_{r_{\mathrm{A}}}^{(\mathrm{OFC})} + \gamma_{\mathrm{A}}^{(\mathrm{Amy}\leftarrow\circledcirc)}\mathbf{a}_{r_{\mathrm{A}}}^{(\mathrm{Amy}\leftarrow\circledcirc)}\cdot\mathbf{v}_{r_{\mathrm{A}}}^{(\circledcirc)} + \gamma_{\mathrm{E}}^{(\mathrm{Amy})}\mathbf{e}_{r_{\mathrm{E}}}^{(\mathrm{Amy})}\cdot\mathbf{x}_{r_{\mathrm{E}}}^{(\mathrm{Amy})} - \gamma_{\mathrm{H}}^{(\mathrm{Amy})}\mathbf{h}_{r_{\mathrm{H}}}^{(\mathrm{Amy})}\cdot\mathbf{x}_{r_{\mathrm{H}}}^{(\mathrm{Amy})}\Bigg) \quad (8.8)$$

The afferent signals $\mathbf{v}^{(\mathrm{OFC})}$ come from Eq. (8.3), while $\mathbf{v}^{(\circledcirc)}$ is a direct reading of the face from the visual afferents in the thalamus, delayed in time with respect to the ordinary visual scene. The activation given from Eq. (8.8) will loop inside the vmPFC by Eq. (8.7).

### *8.2.4 Stealing Is (Conceptually) Wrong*

The artificial moral brain architecture that has just been described is exposed to a series of situations that simulate highly simplified contexts, and can choose between different actions. Some actions are charged with an important survival reward, but in some cases, may cause detriment to others, whose angry reaction will lead to learning that that action is "wrong".

The main input to the model is a visual scene, where three types of objects can appear, at random positions. There is a kind of object with an $X$ shape, and another with a $+$ shape, which are neutral for the individual fitted with the model mini-brain. Only one object is edible, the one with a spherical shape, it might be an apple, an example of a renowned prohibition. Examples are shown in Fig. 8.2. Our artificial subject is unfamiliar with the objects, it can realize how pleasant fruits are to eat, thanks to its taste perception. This sensorial input is simply a matrix $2 \times 2$, in which the ratio of the upper row to the lower row signals how pleasant the taste is. To follow the biblical resemblance, there is an Eden, in this case, simply one quadrant of the overall scene. It is forbidden to eat apples in the bottom right quadrant. In a more profane context, fruits in this quadrant may belong to a member of the social group, and to collect these fruits would be a violation of her property, and would trigger an immediate reaction of sadness or anger. This reaction is perceived in the form of a face with a marked emotion, as the one in the rightmost position, in the bottom row, in Fig. 8.2

After a preliminary phase of maturation of the visual system, like that described in Sect. 6.1.2, the model learns that in the world there are useless objects, like the $X$ shaped and $+$ shaped, but also apples that are delicious to eat. This knowledge becomes coded in the OFC, VS, MD, and vmPFC connections, given in Eqs. (8.3), (8.5), (8.6), and (8.7). This set of equations is an implicit reinforcement learning, where the reward is not imposed externally, but acquired by the OFC map, through its taste sensorial input. No moral norm is yet introduced.

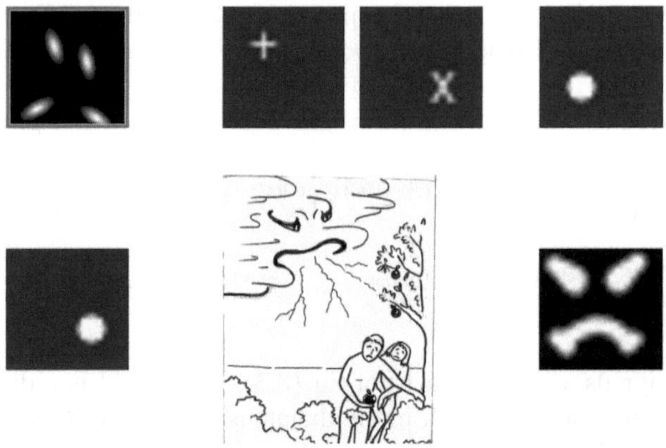

**Fig. 8.2** The visual inputs of the model. In the *top row*, from the left to the right: synthetic elongated blobs, a + shaped and a *X* shaped neutral objects, an edible object, possibly an apple in a free area. In the *bottom row* on the left the edible object is located in the forbidden area (the bottom right quadrant of the scene), corresponding to a sort of Eden where apples cannot be collected (*middle figure*), otherwise a sad and angry schematic face (*right picture*) will appear

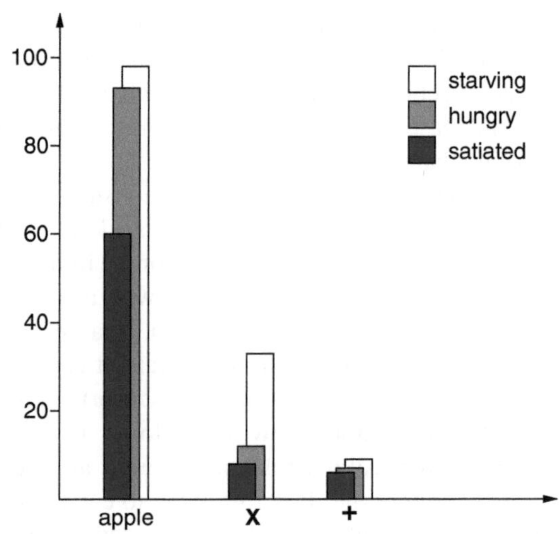


**Fig. 8.3** Fraction of grasping actions selected by the vmPFC model map, depending on the object seen. There is no moral conditioning yet, and the model is tested with three different levels of simulated hunger

The coding in the vmPFC model map corresponds to the decision to grasp or not to grasp the object, and is analyzed with population coding (see Sect. 4.3), further mathematical details are in Plebe (2016). The percentage of decisions to grasp, for each type of object, at three different levels of the parameter $\gamma_{\mathrm{B}}^{(\mathrm{OFC}\leftarrow\circledast)}$ of Eq. (8.3), is shown in Fig. 8.3. When the object is an apple, grasping is always the prevailing choice, that drops to 60 % only in the satiated condition. Occurrence of grasping is instead meaningless for the other objects, except when starving. In this situation, the model decides to grasp *X* shaped objects about 30 % times, and the + shaped objects about 10 % times, even if these objects do not reward.

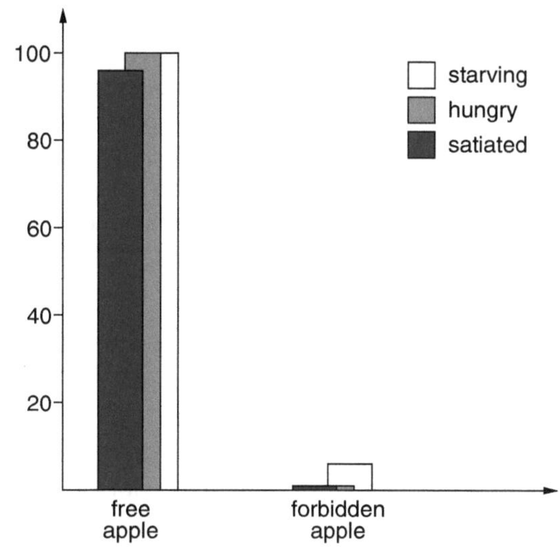


**Fig. 8.4** Fraction of grasping actions selected by the vmPFC model map, depending on the location of the edible object: free or in the forbidden area, after moral conditioning. The model is tested with three different levels of simulated hunger

Soon the model will experience new situations that lead to moral emotion learning, with the objects as stimuli, followed by an image in which there could be an angry face. This face will pop up only when an object of the first kind, the apple, appears in the right bottom quadrant in the scene, the Eden world. This is a sort of private property, and the owner reacts with sadness and anger when his fruit has been grasped. Now the amygdala gets inputs from both the OFC map and directly from the thalamus, when the angry face appears, as from Eq. (8.8), and learns its connections. In this case, there is an implicit reinforcement learning as well, with the negative reward embedded in the input projections to the amygdala.

In Fig. 8.4 there are the percentages of decisions to grasp an object, decoded as before from the vmPFC map. In this case, the samples of the edible object have been divided in two groups, depending on the position in the scene. We can see how strong the inhibition to grasp the edible objects is when placed in the forbidden sector. In both the conditions of normal hunger or satiation, not one grasping decision is made for forbidden apples, while the same fruit in the free territory is grasped 96 % of the times when satiated, and 100 % when hungry. Only under the extreme starving condition are there limited cases of transgression, 0.6 % of the times. It can be claimed that the model has learned a moral rule, as an imperative inhibition to perform certain actions.

One may object that refraining from stealing fruits is nothing more than acting in such a way as to avoid a social negative reaction, learned by reinforcement, which is different from the concept of "wrong". One one hand, we maintain that social reprobation is the standard way of learning moral norms. We agree that there is difference between a pure evaluation of positive and negative rewards, and a moral norm, but this difference, we deem, is in the emotional correlate of the social negative reaction. It should be added that the traditional distinction conventional/moral transgression, has proven to be controversial (Kelly et al. 2007).

On the other hand, we admit that even if the inhibition of stealing in this model can be construed as moral, it is a modest content for the concept of "wrong". The challenge would be to design a model that learns several moral behaviors, and to check if there is a common representation in the maps, that can be a candidate for a general concept of "wrong". We suspect that, in cases of moral violation across very different domains, we would not find a unifying concept, and that what makes morality a single domain for us, is language. But this leads us to the next section.

## 8.3 A Model of the Emergence of "Wrong" in the Brain

This model is a first attempt to derive a possible semantics of moral terms, and in particular, of terms expressing disapproval of a certain action, from the brain representation of the moral violation given by that action. In addition to the theoretical positions on which the model stands, concerning the relationship between moral norms, decisions, and emotions, seen in Sect. 8.2, there is an additional issue concerning the way a term such as "wrong" relates with the mental representations of the individual moral system.

### *8.3.1 The Context of "Wrong"*

As disclosed in Sect. 8.1.1, all the many efforts of including morality within formal semantics, engender a number of problems, most of them related to the need to reconcile the apparent objectivity of a moral statement, with the strong dependency of its truth value on the speaker, his culture, and his individual moral assets. There have been interesting attempts to address this problem by framing it within the more general analysis of words whose meaning is compellingly dependent on the context (Wilson and Carston 2007). Several other classes of terms are affected by similar phenomena, and several of them have been taken as an analogy for moral terms, one is the class of color adjectives, the other, indexicals.

The comparison between moral and color terms can be traced back to Hume (1751), for whom moral values were the "staining of natural objects with the colors borrowed from internal sentiments". This view has been renewed by Wiggins (1987), among others, but dismissed by Blackburn (1985), with a series of objections that apparently settled the color analogy issue. Currently, the analogy with indexicals seems to enjoy more success. It is spelled out more thoroughly by Dreier (1990), and draws on Kaplan (1989)'s standard account of indexicals, according to which part of meaning is dependent on context. In the case of morals, "context" can be defined by the speaker's moral system. This interpretation may fall prey to the more general attack waged against context-dependent semantics put forth by Cappelen and Lepore (2005), but Prinz (2008) argues that moral sentences can survive all their objections.

Our impression is that there is a fundamental agreement in the way both moral and color terms relate with context, which is different from that of indexicals. Speakers, by using indexicals, overtly signal the fact that a significant part of what they want to communicate in a given sentence needs to be gathered from the context. On the contrary, the speaker's assumption in the common usage of color and moral terms is a naive objectivity, and the contextual effect, even if largely in place, is implicitly denied. In addition, we will contend that several of the disanalogies between color and moral terms, found by critics like Blackburn, are based on a universalistic view of color terms, whose weaknesses have been extensively examined in Sect. 7. However, while in the case of color terms there is enough ground upon which to build models reflecting different cultural influences, like for Berinmo, Himba, and English native speakers (see Sect. 7.2), moral modeling is still at a far too sketchy and immature stage to account for such a detailed simulation of cultural variations. Therefore, we will not look to further extend the controversial analogy with other classes of context-dependent linguistic terms. Suffice it to say, that words like "wrong" have the double face of appearing as objective for the speaker, while their semantics is grounded on the context defined by the speaker's moral system, which in turn, in our modeling, derives from emotionally learned norms of behavior. Moral objectivism can still be saved, in case all learned norms in every culture can be shown to derive from shared moral universals, sure not an easy job.

### 8.3.2 Stealing Is (Semantically) Wrong

We presented in Sect. 8.2.4 a neurocomputational simulation of how, in a highly simplified world, the concept that stealing is a moral violation can be acquired. Now the same model is extended with a linguistic component, simulating the emergence of the meaning of "wrong", whose content points to the emotional reaction associated to the possibility of performing the action of stealing. The overall architecture of the model is shown in Fig. 8.5. The components LGN, V1, OFC, VS, MD, Amyg, and vmPFC, are exactly the same as those shown in Fig. 8.1, and described in Sect. 8.2.3. The additional linguistic component is based on the same auditory pathway through Cochlea, MGN, A1, and STS, used in the model visible in Fig. 6.1, and described in Sect. 6.1.1.

The two maps vmPFC and STS project to a single map called PFC (Pre-Frontal Cortex), which, exactly as in Eq. (6.14), is a dramatic simplification of binding semantic coding, which likely spreads throughout large parts of the brain, in a single map. The equation of this map is the following:

$$x^{\text{PFC}} = f\Bigg(\gamma_{\text{A}}^{\text{PFC}}\left(\mathbf{a}_{r_{\text{A}}}^{\text{PFC}\leftarrow\text{vmPFC}}\cdot\mathbf{x}_{r_{\text{A}}}^{\text{vmPFC}} + \mathbf{a}_{r_{\text{A}}}^{\text{PFC}\leftarrow\text{STS}}\cdot\mathbf{x}_{r_{\text{A}}}^{\text{STS}}\right) + \gamma_{\text{E}}^{\text{PFC}}\mathbf{e}_{r_{\text{E}}}^{\text{PFC}}\cdot\mathbf{x}_{r_{\text{E}}}^{\text{PFC}} - \gamma_{\text{I}}^{\text{PFC}}\mathbf{i}_{r_{\text{I}}}^{\text{PFC}}\cdot\mathbf{x}_{r_{\text{I}}}^{\text{PFC}}\Bigg) \tag{8.9}$$

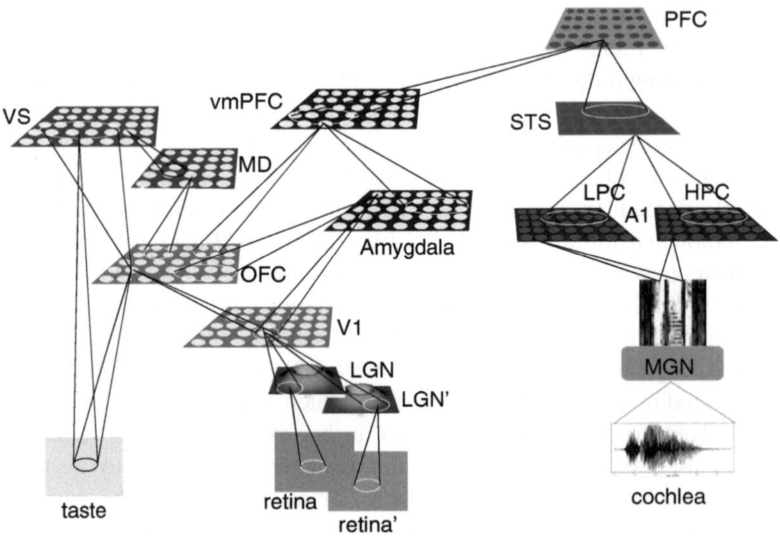


**Fig. 8.5** Overall scheme of the moral semantic model. See Fig. 8.1 for the moral components, on the left of the scheme, and Fig. 6.1 for the linguistic components, on the right

This model, like the previous one, progresses through all the world experiences, and comes to know that among the existing objects there is one that tastes good, but also that stealing it in private orchards has bad consequences. At the same time, its linguistic component will also develop, using as auditory stimuli synthesized waves of the 7200 most common English words with length in range from 3 to 10 characters, like those already used in other models (see Sect. 6.1.2).

At this point a final stage of development occurs, during which every occurrence of a tempting apple in the private sector is associated with hearing the word `wrong`. When the apple is presented in a location where grasping is allowed, the word `good` is associated. Two different conditions of development have been experimented for this last stage: including the appearance of the angry face when the apple is in the private sector, and without the angry face. The former replicates the same condition applied during the moral concept acquisition, based on the emotional negative reinforcement, while the latter only relies on the already established link between negative emotion and the possible moral violation.

The analysis of the content of PFC, and the possible achievement of a semantic coding of "wrong", is performed, as previously, by using population coding, using the algorithm described in Sect. 4.3. The specialization of (4.15) in this case is the following:

$$x_i^{\text{PFC}}(c) \; : \; C \in \mathscr{C} \to \mathbb{R}, \tag{8.10}$$

$$c = \langle s, f, n \rangle \in C = \left( \bigcup_{S \in \mathscr{S}_C} S \right) \times \left( \{\epsilon\} \cup \bigcup_{F \in \mathscr{F}_C} F \right) \times (\{\epsilon\} \cup N_C), \tag{8.11}$$

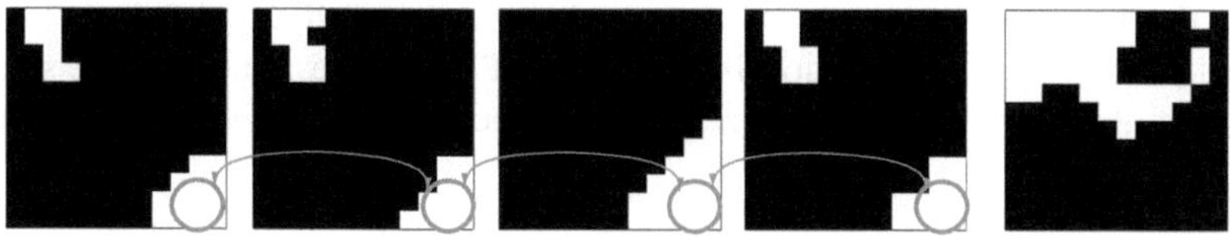

**Fig. 8.6** Population coding of "wrong" in the model PFC map, for several conditions of experiences and development. From left to right: experiences including seeing the angry face and hearing the word "wrong"; without seeing the angry face; without seeing the angry face but hearing the word; including seeing the angry face and hearing the word on a model developed without seeing the angry face. The *gray circles* mark the common semantic code. The rightmost plot is the model PFC map in which are shown all units that are active in the allowed cases, shown to verify that the alleged units for "wrong" get never active in the contrasted situations

where $x_i^{\mathrm{PFC}}(c)$ is computed by (8.9), when the sound $n$ is presented to the auditory path, and in coincidence the scene $s$ is presented to the visual path, followed by face $f$. The kind of objects and, in the case of apples, their position in the scene, introduce a partition in the set of the scene $\mathscr{S}$, such that all sets of variations in the partition $S \in \mathscr{S}_C$ are of that moral category $C$, with $C = \{\texttt{good}, \texttt{wrong}\}$. $N_C$ is the set of utterance naming category $C$. Note that the empty sample $\epsilon$ is included, for experiments in which less stimuli are presented. More precisely, $c = \langle s, \epsilon, n\rangle$ is the case when no face will appear, and $c = \langle s, \epsilon, \epsilon\rangle$ is the case of visual input only, without naming and face reaction.

Figure 8.6 reports the coding found in PFC, for the category `wrong`, under different conditions. The three plots on the left are all obtained with the model developed in the last phase using angry faces as well, the differences are give by the ways of testing. The leftmost picture corresponds to the tests including angry faces, as well as the `wrong` utterance, in the case of apples in the forbidden locations. The next picture moving to the right is the coding testing without faces, with $c = \langle s, \epsilon, n\rangle$. Next to the right is the PFC coding when tested with visual input only, without naming and face, using $c = \langle s, \epsilon, \epsilon\rangle$. The rightmost picture is the comparison with the model developed without ever seeing an angry face reaction, but tested with the full set of stimuli.

A first comment is the strong similarity between the leftmost and the fourth (from left) pictures, which means that for learning the semantic coding the extant stimulus that originally induced a negative emotion is not necessary. The disposition to the same emotion is enough. This fact is confirmed by the similarity between the two pictures on the left. In this case the development phase of the model is the same, but the analysis of the coding is done testing with or without angry faces. The third picture from the left, shows the units in PFC that are activated just by looking at a scene that is reminiscent of a potential violation, that of stealing, without the intervention of a voice saying "wrong", or a face expressing sadness or anger. Still, the coding is provided by a large portion of the same units activated in all the other cases. What is missed is the small number of units segregated in the top left, in all the other cases. Very likely, those units reflect the phonological form of the word `wrong`. It follows that it is possible to consistently identify a region in the model PFC, corresponding to a semantic coding of the term `wrong`, this region has been marked with a gray circle in Fig. 8.6.

## References

Barto, A., & Sutton, R. (1982). Simulation of anticipatory responses in classical conditioning by a neuron-like adaptive element. *Behavioral and Brain Science, 4*, 221–234.

Barto, A., Sutton, R., & Anderson, C. (1983). Neuronlike adaptive elements that can solve difficult learning control problems. *IEEE Transactions on Systems, Man and Cybernetics, 13*, 834–846.

Bechara, A., Damasio, A. R., Damasio, H. R., & Anderson, S. W. (1994). Insensitivity to future consequences following damage to human prefrontal cortex. *Cognition, 50*, 7–15.

Blackburn, S. (1985). Errors and the phenomenology of value. In T. Honderich (Ed.) *Morality and objectivity*. London: Routledge.

Blackburn, S. (1988). *Spreading the word*. Oxford: Oxford University Press.

Blackburn, S. (1998). *Ruling passions – A theory of practical reasoning*. Oxford: Oxford University Press.

Blair, J. (2007). The amygdala and ventromedial prefrontal cortex in morality and psychopathy. *Trends in Cognitive Sciences, 11*, 387–392.

Blair, J. (2010). Neuroimaging of psychopathy and antisocial behavior: A targeted review. *Current Psychiatry Report, 12*, 76–82.

Boll, S., Gamer, M., Kalisch, R., & Büchel, C. (2011). Processing of facial expressions and their significance for the observer in subregions of the human amygdala. *NeuroImage, 56*, 299–306.

Boorman, E. D., & Noonan, M. P. (2011). Contributions of ventromedial prefrontal and frontal polar cortex to reinforcement learning and value-based choice. In R. B. Mars, J. Sallet, M. F. S. Rushworth, & N. Yeung (Eds.) *Neural basis of motivational and cognitive control* (pp. 55–74). Cambridge: MIT.

Bullock, D., Tan, C. O., & John, Y. J. (2009). Computational perspectives on forebrain microcircuits implicated in reinforcement learning, action selection, and cognitive control. *Neural Networks, 22*, 757–765.

Cameron, D., Payne, K., & Doris, J. M. (2013). Morality in high definition: Emotion differentiation calibrates the influence of incidental disgust on moral judgments. *Journal of Experimental Social Psychology, 49*, 719–725.

Cappelen, H., & Lepore, E. (2005). *Insensitive semantics*. Oxford: Basil Blackwell.

Casebeer, W. D., &Churchland, P. S. (2003). The neural mechanisms of moral cognition: A multiple–aspect approach to moral judgment and decision–making. *Biology and Philosophy, 18*, 169–194.

Churchland, P. S. (2011). *Braintrust – What neuroscience tells Us about morality*. Princeton: Princeton University Press.

Damasio, A. (1994). *Descartes' error: Emotion, reason and the human brain*. New York: Avon Books.

Daw, N. D., Kakade, S., & Dayan, P. (2002). Opponent interactions between serotonin and dopamine. *Neural Networks, 15*, 603–616.

Dayan, P. (2008). Connections between computational and neurobiological perspectives on decision making. *Cognitive, Affective, & Behavioral Neuroscience, 8*, 429–453.

Decety, J., Michalska, K. J., & Kinzler, K. D. (2012). The contribution of emotion and cognition to moral sensitivity: A neurodevelopmental study. *Cerebral Cortex, 22*, 209–220.

Dowty, D. R. (1985). *Natural language parsing*. Cambridge: Cambridge University Press.

Doya, K. (2002). Metalearning and neuromodulation. *Neural Networks, 15*, 495–506.

Dreier, J. (1990). Internalism and speaker relativism. *Ethics, 101*, 6–26.
Dupoux, E., & Jacob, P. (2007a). Response to Dwyer and Hauser: Sounding the retreat? *Trends in Cognitive Sciences, 12*, 2–3.
Dupoux, E., & Jacob, P. (2007b). Universal moral grammar: A critical appraisal. *Trends in Cognitive Sciences, 11*, 373–378.
Dwyer, S. (2008). How not to argue that morality isn't innate: Comments on Prinz. In W. Sinnott-Armstrong (Ed.), *Moral psychology, volume 1: The evolution of morality: adaptations and innateness* (pp. 407–418). Cambridge: MIT.
Foot, P. (1967). The problem of abortion and the doctrine of the double effect. *Oxford Review, 5*, 5–15.
Frank, M. J., & Claus, E. D. (2006). Anatomy of a decision: Striato-orbitofrontal interactions in reinforcement learning, decision making, and reversal. *Psychological Review, 113*, 300–326.
Frank, M. J., Scheres, A., & Sherman, S. J. (2007). Understanding decision-making deficits in neurological conditions: Insights from models of natural action selection. *Philosophical Transactions of the Royal Society B, 362*, 1641–1654.
Frankena, W. (1939). The naturalistic fallacy. *Mind, 48*, 464–477.
Geach, P. T. (1965). Assertion. *The Philosophical Review, 74*, 449–465.
Gibbard, A. (1990). *Wise choices, apt feelings – A theory of normative judgment*. Cambridge: Harvard University.
Gläscher, J., Hampton, A. N., & O'Doherty, J. P. (2009). Determining a role for ventromedial prefrontal cortex in encoding action-based value signals during reward-related decision making. *Cerebral Cortex, 19*, 483–495.
Goldman, A. (1970). *A theory of human action*. Englewood Cliffs: Prentice Hall.
Greene, J. D., & Haidt, J. (2002). How (and where) does moral judgment work? *Trends in Cognitive Sciences, 6*, 517–523.
Greene, J. D., Sommerville, R. B., Nystrom, L. E., Darley, J. M., & Cohen, J. D. (2001). fMRI investigation of emotional engagement in moral judgment. *Science, 293*, 2105–2108.
Haber, S. N. (2011). Neural circuits of reward and decision making: Integrative networks across corticobasal ganglia loops. In R. B. Mars, J. Sallet, M. F. S. Rushworth, & N. Yeung (Eds.), *Neural basis of motivational and cognitive control* (pp. 22–35). Cambridge: MIT.
Hare, R. M. (1952). *The language of morals*. Oxford: Oxford University Press.
Harman, G. (1999). Moral philosophy and linguistics. In K. Brinkmann (Ed.), *Proceedings of the 20th world congress of philosophy, Vol. I: Ethics*. Bowling Green: Philosophy Documentation Center.
Hauser, M. (2006a). The liver and the moral organ. *Social Cognitive and Affective Neuroscience, 1*, 214–220.
Hauser, M. (2006b). *Moral minds: How nature designed our universal sense of right and wrong*. New York: Ecco.
Hernandez, M., Denburg, N. L., & Tranel, D. (2009). A neuropsychological perspective on the role of the prefrontal cortex in reward processing and decision-making. In J. C. Dreher & L. Tremblay (Eds.), *Handbook of reward and decision making* (pp. 291–306). New York: Academic.
Holton, R. (2011). Modeling legal rules. In A. A. Marmor & S. Soames (Eds.), *Philosophical foundations of language in the law*, chap 8. Oxford: Oxford University Press.
Hume, D. (1740). *A treatise of human nature* (Vol. 3). London: Thomas Longman.
Hume, D. (1751). *An enquiry concerning the principles of morals*. London: A. Millar.
Kahneman, D., & Tversky, A. (1979). Prospect theory: An analysis of decisions under risk. *Econometrica, 47*, 313–327.
Kaplan, D. (1989). Demonstratives: An essay on the semantics, logic, metaphysics, and epistemology of demonstratives and other indexicals. In J. Almog, J. Perry, & H. Wettstein (Eds.), *Themes from Kaplan* (pp. 481–563). Oxford: Oxford University Press.
Kelly, D., Stich, S., Haley, K. J., Eng, S. J., & Fessler, D. M. T. (2007). Harm, affect, and the moral/conventional distinction. *Minds and Language, 22*, 117–131.

Litt, A., Eliasmith, C., & Thagard, P. (2006). Why losses loom larger than gains: Modeling neural mechanisms of cognitive-affective interaction. In R. Sun & N. Miyake (Eds.), *Proceedings of the 28th annual meeting of the cognitive science society* (pp. 495–500). Mahwah: Lawrence Erlbaum Associates.

Litt, A., Eliasmith, C., & Thagard, P. (2008). Neural affective decision theory: Choices, brains, and emotions. *Cognitive Systems Research, 9*, 252–273.

Mallon, R. (2008). Reviving Rawls's linguistic analogy inside and out. In W. Sinnott-Armstrong (Ed.), *Moral psychology, volume 2: The cognitive science morality: Intuition and diversity* (pp. 145–155). Cambridge: MIT.

Mikhail, J. (2000). Rawls' linguistic analogy: A study of the "generative grammar" model of moral theory described by John Rawls in *A theory of justice*. PhD thesis, Cornell University.

Mikhail, J. (2009). Moral grammar and intuitive jurisprudence: A formal model of unconscious moral and legal knowledge. In D. Bartels, C. Bauman, L. Skitka, & D. Medin (Eds.), *Moral judgment and decision making* (pp. 27–100). New York: Academic.

Moll, J., Zahn, R., de Oliveira-Souza, R., Krueger, F., & Grafman, J. (2005). The neural basis of human moral cognition. *Nature Reviews Neuroscience, 6*, 799–809.

Moll, J., de Oliveira-Souza, R., Zahn, R., & Grafman, J. (2008). The cognitive neuroscience of moral emotions. In W. Sinnott-Armstrong (Ed), *Moral psychology, volume 3: The neuroscience of morality: Emotion, brain disorders, and development* (pp. 1–18). Cambridge: MIT.

Moore, G. E. (1903). *Principia ethica*. Cambridge: Cambridge University Press.

Nichols, S. (2004). *Sentimental rules: On the natural foundations of moral judgment*. Oxford: Oxford University Press.

Nichols, S. (2005). Innateness and moral psychology. In P. Carruthers, S. Laurence, & S. Stich (Eds), *The innate mind – Structure and contents* (pp. 223–242). Oxford: Oxford University Press.

Parkinson, C., Sinnott-Armstrong, W., Koralus, P. E., Mendelovici, A., McGeer, V., & Wheatley, T. (2011). Is morality unified? Evidence that distinct neural systems underlie moral judgments of harm, dishonesty, and disgust. *Journal of Cognitive Neuroscience, 23*, 3162–3180.

Plebe, A. (2016). Neurocomputational model of moral behaviour. *Biological Cybernetics, 109*, 685–699.

Prehn, K., & Heekeren, H. R. (2009). Moral judgment and the brain: A functional approach to the question of emotion and cognition in moral judgment integrating psychology, neuroscience and evolutionary biology. In J. Verplaetse, J. D. Schrijver, S. Vanneste, & J. Braeckman (Eds.), *The moral brain essays on the evolutionary and neuroscientific aspects of morality*. Berlin: Springer.

Prinz, J. (2006a). The emotional basis of moral judgments. *Philosophical Explorations, 9*, 29–43.

Prinz, J. (2008). *The emotional construction of morals*. Oxford: Oxford University Press.

Roedder, E., & Harman, G. (2010). Linguistics and moral theory. In J. M. Doris (Ed.), *The moral psychology handbook* (pp. 273–296). Oxford: Oxford University Press

Rolls, E. (2004). The functions of the orbitofrontal cortex. *Biological Cybernetics, 55*, 11–29.

Rolls, E., Critchley, H., Mason, R., & Wakeman, E. A. (1996). Orbitofrontal cortex neurons: Role in olfactory and visual association learning. *Journal of Neurophysiology, 75*, 1970–1981.

Rolls, E., Critchley, H., Browning, A. S., & Inoue, K. (2006). Face-selective and auditory neurons in the primate orbitofrontal cortex. *Experimental Brain Research, 170*, 74–87.

Rozin, P., Lowery, L., Haidt, J., & Imada, S. (1999). The cad triad hypothesis: A mapping between three moral emotions (contempt, anger, disgust) and three moral codes (community, autonomy, divinity). *Journal of Personality and Social Psychology, 76*, 574–586.

Schnall, S., Haidt, J., Clore, G. L., & Jordan, A. H. (2008). Disgust as embodied moral judgment. *Personality and Social Psychology Bulletin, 34*, 1096–1109.

Sterelny, K. (2010). Moral nativism: A sceptical response. *Minds and Language, 25*, 279–297.

Stich, S. P. (2006). Is morality an elegant machine or a kludge? *Journal of Cognition and Culture, 6*, 181–189.

Suhler, C. L., & Churchland, P. S. (2011). Can innate, modular "foundations" explain morality? challenges for haidts moral foundations theory. *Journal of Cognitive Neuroscience, 23*, 2103–2116.

Thagard, P., & Aubie, B. (2008). Emotional consciousness: A neural model of how cognitive appraisal and somatic perception interact to produce qualitative experience. *Consciousness and Cognition, 17*, 811–834.

Thompson, J. J. (1985). The trolley problem. *Yale Law Journal, 94*, 1395–1415.

Von Wright, G. H. (1951). Deontic logic. *Mind, 60*, 1–15.

Wagar, B. M., & Thagard, P. (2004). Spiking Phineas Gage: A neurocomputational theory of cognitive-affective integration in decision making. *Psychological Review, 111*, 67–79.

Wedgwood, R. (2007). *The nature of normativity*. Oxford: Oxford University Press.

Wiggins, D. (1987). *Needs, values, truth – Essays in the philosphy of value*. Oxford: Clarendon.

Wilson, D., Carston, R. (2007). Word meanings in context: A unitary relevance-theoretic account. In M. Aurnague, K. Korta, J. Larrazabal (Eds.), *Language, representation, and reasoning: Memorial volume to Isabel Gómez Txurruka* (pp. 283–313). San Sebastián: University of the Basque Country Press.

Young, L., & Dungan, J. (2012). Where in the brain is morality? Everywhere and maybe nowhere. *Social Neuroscience, 7*, 1–10.

# Chapter 9
# Semantics: What Else?

**Abstract** The previous chapters have given samples of neurosemantics addressing specific semantic phenomena, using a unified neurocomputational approach. Much of the complexity of real language has been neglected, and in this chapter other developments will be presented, which fill in some of gaps that remain. The models presented in this chapter are not original developments of the authors, their selection is due to their theoretical grounds and their motivations, which are perfectly in line with the neurosemantics enterprise, as we have defined it here. In particular, Friedemann Pulvermüller and his associates have attempted to answer questions in semantics by developing neurocomputational models, based on brain representational mechanisms, as those here described in Chap. 3, and compatible with the organization of brain areas involved in language processing.

Section 7.1 has shown how naturally the first instance of syntax emerges from sequences of adjectives and nouns. From that first step into a complete management of syntax, the brain needs to organize circuits to handle higher order combinatorial information, something simulated by Pulvermüller and called discrete combinatorial neuronal assemblies. With this computational architecture it is possible to explain main syntactic structure, such as that of verbal phrases.

Much more is needed, but what has been achieved so far, the mathematical frameworks laid down, the definition of the research project, strongly suggest that neurosemantics today is feasible, and is one of the deeper and most appropriate efforts in explaining linguistic meaning.

## 9.1 Neurons, Word Order and Verbs

Most of the neurosemantic models discussed in the previous chapters simulated phenomena at a lexical level. The only exception was the model of nouns and adjectives (see Sect. 7.1), a step towards what we called an "embryonic-syntax". From there to full blown language there is clearly a long way to go. Even if the cognitive semantics enterprise (see Sect. 5.2) has drastically mitigated the role assumed by syntax in early days of cognitive science, there is no doubt that the rules and principles that govern the sentence structure of any given language are a fundamental component of language meaning. Moreover, our simulations did not cover all grammatical categories, with verbs, one of the most important, not having

 
A. Plebe, V.M. De La Cruz, *Neurosemantics*,
Studies in Brain and Mind 10, DOI 10.1007/978-3-319-28552-8_9

been addressed. This section will concentrate on the research and results of a group that, in our opinion, is currently offering the most advanced insights on syntax using a perspective that is in line with our neurosemantics view.

### 9.1.1 The Brain Bases of Syntax: Exploring the Mechanics

Friedemann Pulvermüller and his associates have proposed a mechanism called discrete neuronal binding or discrete combinatorial neuronal assemblies (DCNAs) to explain how syntax theories, might be grounded in neuronal circuits and synaptic learning. They use brain inspired artificial network models with strong auto-associative links, to show that through Hebbian learning, their model can learn discrete neuronal-representations that can function as a basis for syntactic rule applications and generalization (Pulvermüller and Knoblauch 2009).

The approach adopted by this group, lies in the middle, or perhaps more accurately "bridges", the views and research practices often adopted by a large number of neural network modelers investigating natural language processing, and that of a consistent group of formal linguists. The view held by the former, is that there is no need to invoke rules in the processing of language in the brain, since probability mapping in simple artificial recurrent networks is enough for simulating the emergence of syntax in artificial models (Rumelhart and McClelland 1986a; Elman 1990). This would be due to the fact that the brain is an organ that is extremely sensitive to regularities in the environment, and thus, the brain of a typical language learner would naturally also pick up on the syntactic regularities in the linguistic input and learn the statistical regularities with which words appear together in her language. The well-known view of the latter, is that syntactic representations are the fruit of the processes of dedicated innate mental machinery (Chomsky 1964, 1966). For one, learning is of paramount importance, and in particular associative learning processes, for the other, it is only a secondary phenomenon.

Building on a vast amount of evidence made available by contemporary neuroscience, the concept of discrete neuronal binding or discrete combinatorial neuronal assemblies (DCNAs) proposes a new take on the debate that targets recent evidence on neural wiring in the brain, and the role that the associative learning of combinatorial information that is inherent to word strings, might play. DCNAs are artificial networks that incorporate auto- and hetero-associative connections for regulating excitation that simulate those found in the cortex. The networks also have neuronal devices for sequence detection built in, justified by evidence of the existence of neurons that respond to input patterns (or sequences) found in a variety of animals. These researchers thus consider it likely that sequence detectors are also found in the human brain and might be involved in the processing of combinatorial information in sentences (Pulvermüller 2002a,b). The most elementary circuit implementig this detector is composed by two input units $\alpha$ and $\beta$ connected to a third unit $\gamma$, that becomes active only if the activation of $\alpha$ is followed by that of $\beta$, within a certain time windows. The inverted sequence of activation $\beta \rightarrow \alpha$ will not activate $\gamma$.

These detectors would provide the mechanism for dynamically linking constituents, realizing syntactic links. The linkage would include, for example, "the previous activation of the first and the second word in a string (e.g. noun-verb), the resultant ignition of the order-sensitive DCNA and a binding process functionally linking the latter with the active lexical representations" (Pulvermüller 2010). The binding mechanism here is conceptualized as being synchronized oscillatory neuronal activity at high frequencies. These oscillatory dynamics are considered by a number of researchers to reflect lexical and sentence processing. The authors report that "as the networks map coincident neuronal activation driven by the co-occurrence and substitution patterns of string segments in sentences, they 'grow' putative network equivalents of discrete rules or discrete combinatorial neuronal assemblies" (Pulvermüller and Knoblauch 2009).

The proposed mechanism by which combinatorial neuronal assemblies would emerge are the following (Pulvermüller 2010).

a. Elementary sequence detectors sensitive to ordered pairings (e.g. words/morphemes), through reciprocal connections and synchronization, would functionally link or merge with two lexical circuits, thus building a higher-order syntactic unit. Important here, is that a range of binding units for word pairs become linked among each other.
b. This emerging group of aggregates of sequence detectors, as a result, becomes sensitive to any lexical element that is part of a particular syntactic class or lexical category followed by any other element belonging to a different lexical category.

Below, is a brief summary of the neurocomputational simulation experiments this group has done and the results obtained. In their simulation experiments on syntactic learning Pulvermüller and Knoblauch (2009) used a pre-structured auto-associative memory with built in sequence detectors for every possible pair sequence of words. The grammar area of the network included pre-wired sequence detectors, as well as initially weak links between all pairs of sequence detectors, in keeping with an important feature of the cortex. In the model, the availability of sequence detectors and the auto-associative links between them resulted in the binding of the sequence detectors into circuits operating on classes of lexical items.

In Fig. 9.1 we can see how the sequence detectors found in the grammar area of the network formed strong connections to their corresponding lexical representations (red lines to gray dots at left and top), and among each other as well (black lines). The connections between the elementary sequence detectors that were sensitive to word couplings were selectively strengthened. This was due to the regular co-activation of sequence detectors during the process of string learning and word substitutions between strings. It was this co-activation of sequence detectors that led to the development of neuronal aggregates including sequence detectors sensitive to similar contexts. According to the authors it is thanks to these strong internal connections within neuronal aggregates, that they

| | sass | sprach | mach | leg | sag | klirr | brann | weh | zurueck ging | gut tun |
|---|---|---|---|---|---|---|---|---|---|---|
| Hahn | x | x | x | | | | | | | |
| Esel | | x | x | x | x | | | | | |
| Hund | x | | | x | x | | | | | |
| Katze | x | | x | | | | | | | |
| Raeuber | x | x | | | | | | | | |
| Scheibe | | | | | | x | | | | |
| Licht | | | | | | | x | | | |
| Wind | | | | | | | | x | | |
| Kraefte | | | | | | | | | x | |
| Knochen | | | | | | | | | | x |

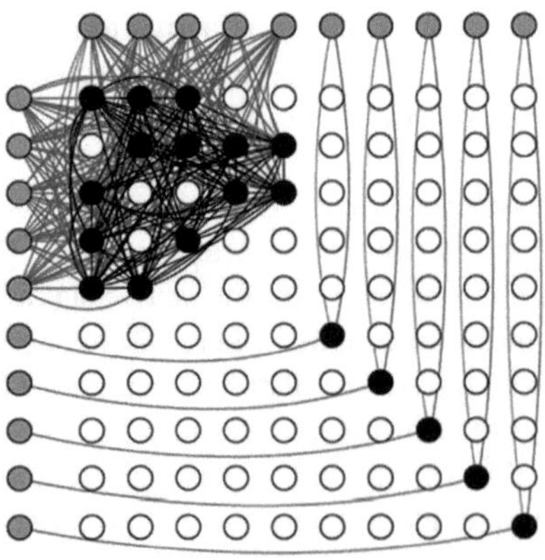

**Fig. 9.1** Combinatorial information found in noun-verb sequences and network result of learning this information in an auto-associative memory including sequence detectors for word pairs. *On the left*: matrix of co-occurrences and substitutions of 20 nouns and verbs from British National Corpus. *On the right*: network of lexical circuits and sequence detectors and the connections strengthened by learning the string set from BNC. *Gray circles* are neural units corresponding to words in diagram on the left. Central matrix shows sequence detectors corresponding to word pair sequences. *Black circles* indicate sequence detectors whose associated word pair was in the input, leading to strengthening of connections between sequence detector and word representations (*light lines*). *Dark lines* show strengthened auto-associative links between pairs of sequence detectors. At top left and bottom right, discrete combinatorial neuronal assemblies, DCNAs, have formed. Depending on the threshold of activation, these DCNAs either bind all nouns to verbs, or provide specific syntactic-semantic linkage of action verbs and nouns related to living entities, and of flight-related verbs and flying-object nouns (Adapted from Pulvermüller and Knoblauch 2009)

were able to act as a higher-order discrete functional unit for syntactic binding. In addition, generalization in the network is explained through the use of a very basic mechanism. A possible first word is bound into the syntactic binding circuit, just as any possible second word is. As a result, activity spreads from any active neuronal unit that is part of the first lexical class to all the items that are part of the second class, even if the sequence has not been encountered before. The authors argue that, "after learning, syntactic binding units formed in the network provide a mechanistic correlate of an abstract binary rule" (Pulvermüller 2010).

What makes this model different? According to these authors, it is the neurophysiological and neuroanatomical features of the brain they have incorporated in their model, that makes their model stand apart from many of the artificial neural networks most commonly used to investigate language. These are the following:

1. Reciprocal auto-associative connectivity
2. Built-in elementary sequence detectors
3. Unsupervised Hebbian-learning
4. Sparse coding
5. Inhibitory circuits

Using these features, their model simulates the learning processes brought about by strings with multiple mutual substitutions between them and the emergence of possible neuronal correlates of the binary rules proposed by a number of grammar theories.

We would like to comment here on how the achievements of Pulvermüller and his collaborators relate to our models. The commitment, in principle, is much the same, that of explaining crucial phenomena of human linguistic performance neuromechanistically. Several of the brain mechanisms they use are the same: Hebbian learning, reciprocal excitatory and inhibitory connections, sparse coding. For what concerns the linguistic phenomena they target, Pulvermüller and his collaborators venture into territories we have not: noun-verb sequences, the only syntactic structure we explore is the more limited adjective-verb (see Sect. 7.1). However, there has been a price they have necessarily paid for such a bold enterprise. We have been able to simulate, even in a very simplified way, the entire path from the external world to internal neural representations of linguistic entities, avoiding any arbitrary coding of linguistic elements of object features as neural inputs. All the external exchanges of our models are through their sensorial pathways, simulated as closely as possible according to what is known regarding brain organization. On the contrary, the DCNAs structures that realize noun-verb comprehension are stand-alone models, detached from the rest of the brain, and their inputs are pre-coded linguistic elements. This move has several advantages, for example, the model can learn directly from large text corpora (Pulvermüller 2010), and use words like `start` or `hate`, which would be extremely difficult to teach a model using simulated perceptual experiences. We think a great challenge for the future development of neurosemantics from a mechanistic perspective, will be the integration of models that rely purely on sensorial experience, with neural components able to manage complex syntactic structures, like DCNAs.

### *9.1.2 Words and Action Perception Circuits*

As the section above illustrates, words are not isolated entities, and our attribution of meaning to them does not take place in a vacuum. Words are used in relation with other words, and their meaning is influenced and modified by the way these words are combined in language. This combination is governed by the rules that regulate how they can be combined in a particular language. This "word-word" combinatorial knowledge is essential to the construction of our semantic system. We use words in a world, however, and how the words we use, refer to and are grounded in our experiences and interactions with the world, or "word-world" knowledge, also needs to be incorporated in the meaning we attribute to at least a portion of the words we learn (Cangelosi et al. 2001; Pulvermüller 2013). For this to be possible, links between words, objects, their sensorimotor features and actions need to be established. To address how different aspects of "word-word" as well as "word-world" knowledge is acquired, and how it is "brain-embodied", Pulvermüller has also proposed the Action Perception Theory of Semantic Circuits

(APT). It is a cognitive model with a strong biological basis, that is related to the one discussed in the previous section. It integrates the combinatorial mechanisms described above, and helps to account for how both contribute jointly to semantic learning. It is supported by a body of neuroscientific evidence that indicates that words are organized in the brain as distributed cell assemblies and their meaning is reflected by these cortical distributions. How the meaning of words is represented in the brain depends on how these words are encountered and learned, the correlated motor activity and the interactions the body has with objects, other bodies, and the world. In one neurophysiological and behavioral study, different types of verbs, for example, were found to be processed at different speeds, and to have different cortical topographies (Pulvermüller et al. 2001).

A key concept here, is that "correlated activity in both sensory as well as motor brain systems, and particularly in the cortex, as well as already established neuroanatomical connections drive the creation of significant building blocks of cognition, language and meaning" (Pulvermüller 2013). The formation of functional units, neuronal assemblies or action perception circuits (APCs), would take place thanks to this correlation and would have specific functional properties. Primary motor, sensory areas and visual areas, are not linked directly but rather indirectly in the brain, with reciprocal local connections as well as long-distance connections providing the in-between area connections.

Thanks to learning, sensory and motor circuits can become connected to each other. The mechanisms of the linkage between action and perception circuits, have been explored by this group also with neurocomputational modeling. For example Garagnani et al. (2008) simulated the left-perisylvian language cortex and targeted how the creation of APCs for spoken word forms are created. This model realized important features of connectivity between primary motor and auditory areas and highlighted the role these mechanisms might play in language comprehension and attention processes. In a similar vein, a very recent study using the cell assemblies paradigm, explored the learning of associations between visual and motor modalities in neurorobotic experiments using the iCub cognitive robot (Adams et al. 2014). Results indicate that the learning of robust cell assemblies allows the robot to select a correct motor response based on visual input alone. Further fine tuning of the neural model for learning cell assemblies, showed its potential in being used as a controller for the robot in visuo-motor association tasks. This work extends the exploration of brain embodied aspects of cognition, strongly based on neuroscientific data, also to the artificial agents and robotics domain, which by their very essence are the ideal test beds for embodied cognition theories.

## 9.2 Building a Semantics of Numbers

This section will address the semantics of a peculiar class of words: Number words. How do children learn number words and the concepts related to them? How do they learn what number words mean? How do children interpret them? The questions may seem quite similar, but they concern very different aspects of what is involved in the acquisition of number vocabulary. Psychologists have been interested for quite some time in finding answers to the first question (Gelman and Gallistel 1978; Fuson 1988; Wynn 1992; Dehaene 1997; Carey 2001, 2004). Theoretical linguists have been pondering the second and third issues, in particular, for perhaps even longer, using a variety of theoretical models, and more recently, experimental methods, to arrive at the answer (Horn 1972; Levinson 2000; Geurts 2003; Musolino 2004, 2009). In the last decade, an effort has been made by a small group of scholars from each of the respective fields to integrate some of the evidence that has emerged from their respective studies on how children learn to use number words and arrive at understanding their meaning, as well as how their understanding differs from that of adults (Papafragou and Musolino 2003; Musolino 2004, 2009; Hurewitz et al. 2006; Huang et al. 2012).

Quite recently, embodied cognition researchers have also begun to investigate aspects of these issues, with some very recent attempts employing neurocomputational and developmental robotic models to study how these processes might be bootstrapped in very young children, but based on evidence emerging almost exclusively from developmental psychology. A detailed discussion of all three approaches is beyond the scope of this work, but this section will provide brief sketches of these approaches and in particular, on research investigating the connection between learning the count system along with number words and the implications this learning might have on the child's building an understanding of what number words mean.

## 9.2.1 Learning Number Words: Does Counting Count?

Though children as young as 6 months have been found to be able to discriminate between set sizes, and children 1 and 2 years of age have demonstrated to be good at reciting the count sequence (Gelman and Gallistel 1978; Fuson 1988; Wynn 1992), as well as capable of recognizing number words as designators of quantity (Wynn and Bloom 1997), their difficulty seems to lie in understanding how specific words match to specific quantities.

The pathway by which children acquire number words can be considered as highly indicative of how they learn to associate meaning to these words. Number words are highly frequent in child directed speech, but their meanings are acquired slowly, with effort and in stages. The child's first "hands on" encounter with number words, however, most often comes through learning the counting routine. Gelman and Gallistel (1978) proposed that the representation of integers is part of our innate cognitive endowment and put forth a number of innate principles that would be guiding the child's acquisition of the number vocabulary leading to the transition from being rote repeaters of a verbal sequence to understanding what those number words really mean.

These principles are: (1) the one-one principle (involves the assignment of one and only one distinct counting word to each of the items to be counted), (2) the stable-order principle (when counting, number words are always assigned in the same order), (3) the cardinality principle (the last number word uttered when counting is the total number of objects in a set) (4) the abstraction principle (the preceding principles can be applied to any set of objects in a set, tangible or not, animate or inanimate, etc.), (5) the order irrelevance principle (knowledge that the order in which objects are counted is irrelevant, so long as every item in the set is counted only once).

While there is wide consensus on what may be the guiding principles behind the child's acquisition of a mature counting system, there is less consensus on the extent to which these principles can explain other aspects of development in the number learning domain (Musolino 2009). One view, the "principles before skills" (Gilbert and Wiesel 1983), or the "continuity hypothesis" (Corre and Carey 2007) endorses the nativist thread running through G&G's account concerning the representation of integers and the implicit knowledge of the counting principles.

An alternative view, would be the "skills before principles" view or the "discontinuity hypothesis", which instead sees the representation of integers and the knowledge of the counting principles as an emergent phenomenon, deriving from experience, at least in part (Fuson 1988; Karmiloff-Smith 1992; Wynn 1992; Spelke and Tsivkin 2001). What these scholars propose is that children identify number words and arrive at their meaning thanks to their sensitivity to verbal input, as well as through knowledge of the principles (such as those proposed by Gelman and Gallistel) but they would arrive at this knowledge inductively and relatively late during the pre-school years.

While learning to count might represent an initiation to the use of number words, and to the understanding that numbers are often used to express cardinality, how children in time come to fully understand and interpret these number words is quite another story, a linguistic one. We must not forget that number words are acquired through language and are embedded in and used, in the context of language.

### 9.2.2 Learning About Number Words: What Else Counts?

Investigations in linguistic theory at the interface between (lexical) semantics and pragmatics have also addressed the question of how the understanding and interpretation of number words might unfold in development.

An important argument behind this research, as stated by one of the primary actors involved, the linguist Julien Musolino (2009), is that "linguistic behavior of number words extends far beyond counting and into the realm of syntax, semantics and pragmatics". In psychological studies, it has been generally assumed that number words have exact meanings (e.g. three means EXACTLY THREE), and that subjects in these studies interpret number words as such (Wynn 1992). Linguistic accounts of number semantics, however, argue that number words not only have lower-bounded meanings (e.g. AT LEAST THREE), but that they can also be interpreted in several different ways, depending on the linguistic context. According to these accounts, speakers would restrict their reference through pragmatic inference or scalar implicatures.

Notwithstanding the recent forays of developmental psychology researchers into the linguistic theoretical arena (Hurewitz et al. 2006; Huang et al. 2012), demonstrate the one having gained awarness of the utility of theoretical models and experimentation of the other, and vice versa, the debate on issues like the one above continues. In fact, Huang et al. (2012) found that under their experimental task conditions using a novel paradigm, which teased apart semantic and pragmatic aspects of interpretation, both children and adults consistently gave exact interpretations for number words. They interpret their results as "unambiguous evidence demostrating that number words have exact semantics".

### 9.2.3 The Case of Numerical Quantified Expressions

According to Musolino (2004, 2009), however, any account of how children learn to interpret numerical relations through language is incomplete without consideration of the logico-syntactic properties of number words, for example. On this view, as demonstrated by the example of numerical quantified expressions (NQE), numerals do not only interact with each other (e.g. three boys are holding two balloons) but with other quantified expressions as well (e.g. three boys are holding each balloon). The linguistic complexity underlying the interpretation of NQE, as a case in point, makes understanding how children acquire the facts necessary to their proper interpretation of crucial importance to developmental accounts of language and number cognition. Theoretical linguistic accounts had already posited that NQE could be interpreted in at least four ways. In psycholinguistic experiments with both children and adults, Musolino tested this hypothesis, and in particular, his belief that the various readings or interpretations possible of NQE, are not learned by young children, but are instead implicitly deduced thanks to the combinatorial power of language, and in this particular case, the compositional aspects of semantics. That is, the way that the meaning of a sentence can be systematically deduced from the meaning of its parts. What he predicted was that once children had acquired the meaning of expressions like two N, three N, etc); and had a basic command of the core grammatical principles of their language, as well as understand the compositional aspect of word meanings, they should then be able to implicitly deduce the range of meanings arising from the interaction of multiple NQE.

On a side note, another proposal on how children might be arriving at the knowledge of such complex properties of language, sustains that it might be derived by experience thanks to their sensitivity to the distributional properties in their linguistic experience, rather than by a grammatical or pragmatic competence (Gennari and MacDonald 2005), for a recent related view and review pertaining to infant numerical abilities see Cantrell and Smith (2013).

As has been mentioned above, not all number word learning situations are the same. Numbers and number words appear in linguistic expressions, have multiple senses: they can be arithmetic entities (e.g. indicating exact cardinality of a set) or they can be used as quantifiers (e.g. as in NQE such as two girls, three balloons) and as such are subject to a range of interpretations, and they also interact with other lexical items that are similar syntactically as well as semantically (Syrett et al. 2012). How children master all the different uses and meanings of number words remains an open and challenging research question. Studies such as the ones mentioned above (Musolino 2004, 2009) and examples of more recent work in this direction (Huang et al. 2012; Syrett et al. 2012) have begun to show how the integration of work in theoretical linguistics and developmental psychology addressing questions on the acquisition of number vocabulary, could have significant implications for developmental accounts of how it occurs.

In the section below, we present another relatively new approach to the study of how number words get their meaning, which applies the frameworks of embodied number cognition and developmental cognitive robotics, and targets the role of finger counting as a bootstrapping technique.

### 9.2.4 Embodied Cognition Accounts: The Case of Finger Counting

As we have already mentioned, the pathway by which children acquire number words can be highly indicative of how they learn to associate meaning to these words. Early finger counting is an example of an important strategy children use that might be assisting them in mapping number words onto their meanings. Whether it is an essential stage in the development of this process is highly debated, however, but there is strong evidence on the positive contribution of sensorimotor skills and representation in the development of numerical cognition. A growing number of researchers, claim that finger counting is an important tool children as well as adults use across a variety of cultures in the development of numerical cognition (Andres et al. 2008).

The topic of finger based number knowledge has, in fact, seen a surge of new interest, especially from embodied cognition perspectives, for a special issue on the topic see Fischer et al. (2012). Finger counting has generally been assumed to be important to the acquisition of a mature counting system as well as instrumental to the development of children's arithmetic abilities. It would help children acquire the previously mentioned principles put forth by Gelman and Gallistel (1978), considered to be fundamental to the child's understanding of the counting system. Recent studies have reported an association between finger gnosis (the ability to mentally represent one's fingers) and mathematical abilities (Costa et al. 2011), and found finger training helpful in improving the performance of children with weak numerical skills (Gracia-Bafalluy and Noël 2008). Evidence such as this supports the view that finger representations play a special role in number cognition, and might serve as a basic building block in the child's unfolding capacity to mentally manipulate abstract numerical information.

A close link between finger counting strategies and patterns, and how they may influence the mental representation and processing of number, has also been suggested by evidence coming from neuroimaging studies. Studies using fMRI on adult subjects have found intrinsic functional links between finger counting and number processing. Cortical motor activity is evoked both by Arabic digits and number words, which reflect particular individual finger counting habits (i.e. whether when counting small digits subjects started with their right or left hand) (Tschentscher et al. 2012). One interpretation of these results invokes a shared neural network for number processing and planning of finger movements, which would include parietal cortical areas, the precentral gyrus and the primary motor cortex, in which number perception might very well elicit the sub-threshold tendency to move associated fingers. These authors, explain that the association between numbers, number words and individual finger counting movements might have come about in their subjects during their individual development of numerical skills in childhood, and would be predicted by a Hebbian learning approach to semantic circuits (Pulvermüller 1999). The prediction is, that due to the fact that children often use their fingers when counting and solving simple counting problems, a correlation between the neuronal activation for the processing of numbers and the movement of fingers is established. What Pülvermuller and his collaborators propose here, is compatible with the general approach embraced in this book and in particular, with our account of coincidence detection (see Sect. 3.2).

Developmental as well as neurocognitive studies, in keeping with what has been found in neuroimaging studies, suggest that finger counting activity, helps build motor-based representations of number that continue to influence number processing well into adulthood, suggesting that abstract cognition may be rooted in bodily experience (Domahs et al. 2010). In fact, these motor-based representations have been argued to facilitate the emergence of number concepts from sensorimotor experience through a bottom-up process (Andres et al. 2008).

According to De La Cruz et al. (2014) and Di Nuovo et al. (2014a,b), finger counting, can also be seen as a means by which direct sensory experience with the body can serve the purpose of grounding number as well as number words initially as low level labels, that later serve as the basis for the acquisition of new higher level symbols from the combination of already grounded ones, something known as grounding transfer (Harnad 1990). The grounding approach has also been useful for the modeling of the acquisition of words for objects (Witzel and Gegenfurtner 2011) and for actions as well as for numbers (Rucinski et al. 2012).

As we have discussed in previous sections, learning to count also involves the acquisition and use of a number word system. One of the major questions regarding the acquisition of a number word system is how children come to understand how specific words refer to specific quantities. One proposal already touched upon above, is that the syntax of number words as well as the contexts in which they appear, might be serving as cues that help children bootstrap this process early. According to one group of researchers, syntactic bootstrapping could also be operating in tandem with the counting principles in helping children pick out number words in the linguistic context and arrive at their meaning (Syrett et al. 2012). From a brain point of view, finger counting experience coupled with that of number words may also be serving as an early entry point to this understanding, via the storing of number word meaning through the interlinking of action-perception circuits in the brain, in what is known as correlational learning (Pulvermüller 2013), or what we have referred to in other sections as coincidence learning.

In sum, while finger counting may not be strictly necessary for children to get on their way to the cognition of number, there is evidence that it does seem to help the learning process, serving as a bridge between possibly innate abilities to perceive and respond to numerosity (Butterworth 2005) and the development of the capacity to mentally represent and process number as well as linguistic number related concepts (Lafay et al. 2013).

### *9.2.5 Modeling Finger Counting and Number Word Learning*

A number of connectionist models have simulated different aspects of number learning. Ma and Hirai (1989) for example, studied how children learn to count using an associative memory network model, which mimicked three phenomena proposed by Fuson (1988), to be present in the acquisition of counting by children (i.e., number word sequence produced by children dividable into three distinct portions: conventional, stable nonconventional, and unstable; irregular number words (e.g. "fifteen") omitted more often than regular ones ("fourteen", "sixteen"); initially number word sequence is in recitation form). Ahmad et al. (2002), explored quantification abilities and how they might arise in development, using a multi neural net approach, that combined supervised and un-supervised nets and learning techniques in order to simulate subitization (i.e., phenomenon by which subjects appear to produce immediate quantification judgments, usually involving up to 4 objects, without the need to count them) and counting. They used a combined and modular approach, providing a simulation of different cognitive abilities that might be involved in the cognition of number, (each of which would have their own evolutionary history in the brain). Rajapakse et al. (2005), targeted aspects of language related to number such as linguistic quantifiers. Using a hybrid artificial vision connectionist architecture, they ground linguistic quantifiers such as few, several, many, in perception, taking into consideration contextual factors. Their model, after being trained and tested with experimental data using a dual-route neural network, is able to count objects (fish) in visual scenes and select the quantifier that best describes the scene.

Note that all these models belong to the predictive category seen in Sect. 3.1.5, their components often have no direct correspondence with brain components, and therefore, cannot meet the model-mechanism-mapping criteria (Kaplan 2011; Kaplan and Craver 2011). Not many models, to our knowledge, have attempted to simulate number learning with neurocomputational models, as those used in the neurosemantic simulations of the previous chapters. Recently, an advance in simulating the semantics of numbers has been the use of models that, even if missing a strict correspondence with the brain, are embedded in a physical body, that of robots. Rucinski et al. (2012), using a cognitive robotics paradigm explored embodied aspects of counting, and in particular, the contribution of counting gestures such as pointing. This model, however, did not consider the role of finger counting and the acquisition of number words in numerical abilities.

A very recent approach using a cognitive developmental robotics paradigm (Asada et al. 2009; Cangelosi and Schlesinger 2015) explored whether finger counting and the association of number words (used as tags) to the fingers, could serve to bootstrap the representation of number in a cognitive robot enabling it to perform basic numerical operations, such as addition (De La Cruz et al. 2014; Di Nuovo et al. 2014a,b). The robotic model used for the experiments was a computer simulation model of the iCub humanoid robot (Tikhanoff et al. 2011). The iCub is an open-source humanoid robot platform designed to facilitate cognitive developmental robotics research as described in Metta et al. (2010). At its current state the iCub platform is a child-like humanoid robot 1,05 m tall, with 53 degrees of freedom (DoF) distributed in the head, arms, hands and legs. The simulated iCub has been designed to reproduce, as accurately as possible, the physics and the dynamics of the physical iCub. The simulator allows the creation of realistic physical scenarios in which the robot can interact with a virtual environment. Physical constraints and interactions that occur between the environment and the robot are simulated using a software library that provides an accurate simulation of rigid body dynamics and collisions. The study focuses on the fingers of the robot that has 7 degrees of freedom for each hand. Figure 9.2 shows the finger representations of numbers one through five with the right hand of the robot (numbers from six to ten are represented by adding left hand fingers with all the right hand fingers open).

Figure 9.3 instead, shows the architecture of the robot's cognitive system, in which the different units and their connections are presented in a schematic form. The lower part of the implemented neural system is directly connected with the robotic platform, and can be summarized in: (i) the motor controller/memory (Motor System and Right/Left Layers), that is able to plan finger movements by setting the finger joints' angles and to memorize the finger number sequence; (ii) an

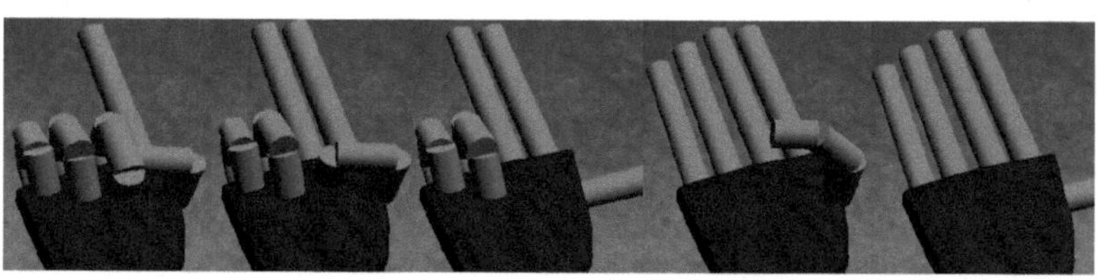

**Fig. 9.2** Number representation with the right hand fingers of the iCub. From left to right: one, two, three, four and five

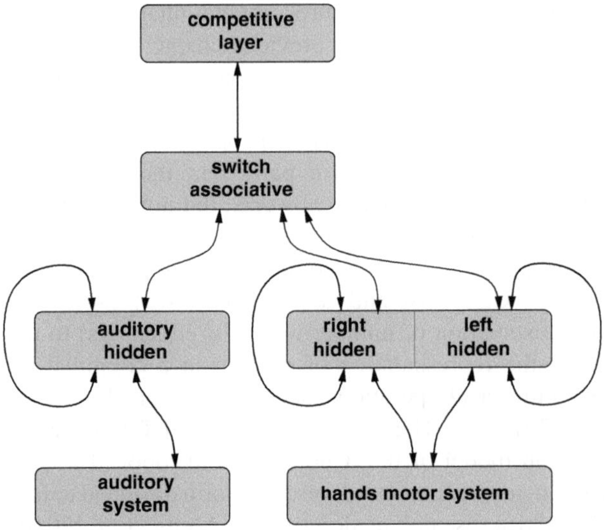


**Fig. 9.3** Schematic of the Robot's Cognitive System. In the lower part are the units of the motor controller/memory and of the auditory sub-systems, they are directly connected with the robotic platform. In the upper part there are the units with abstract functions that are the switch/associative network and the competitive layer classifier. *Bold links* indicate a full (one-to-all) connection between each layer, while dotted links are direct (one-to-one) connections. Note that the system's external inputs coincide with the outputs, indeed proprioceptive information from motor and auditory systems is an input for the system during the training phase, while it is the control output when the system is operating

auditory memory (auditory system and auditory layer), that is able to memorize the number words sequence. The upper part of Fig. 9.3 presents the inner units that are responsible for abstract functions (i.e. not directly connected with the robot), they are the switch/associative layer, that allows the two lower systems to cooperate in order to perform other functions, and the competitive layer classifier we implemented to test the quality of the number learning. After supervised training, it is able to represent the correspondence between numbers from 1 to 10 and the internal representations (i.e. hidden layer activations and/or cepstral coefficients).

The role of the competitive layer classifier is to simulate the final processing of the numbers, after a number is correctly classified into its class, the appropriate action can be started, e.g. the production of the corresponding word, of a symbol, the manipulation of an object and so on. The motor controller/memory is based on the idea of recurrent networks (RNN), introduced by Pearlmutter (1989) and refined by Elman (1991), which is an artificial connectionist network, able to incorporate temporal dynamics in a simple, symbolic way, thanks to feedback connections in the hidden layer units. This controller uses two different RNNs in order to model lateralization when processing numbers, and the network that controls the left hand will be switched off when low numbers (1–5) are processed. The two RNNs that compose the motor controller/memory were trained separately, i.e. with different random weight initialization. The motor controller is implemented by two different RNNs, trained separately, but are referred to as a single unit. The use of RNNs to learn to count was investigated in the recent past by Rodriguez et al. (1999). They explored the capabilities of recurrent networks in the task of learning to predict the next character in a simple deterministic context-free language, in order to provide a more detailed understanding of how dynamics could be harnessed to solve language problems.

The artificial neural networks were implemented using the Matlab Neural Network Toolbox 8.0, the supervised training algorithm for all networks was the Levenberg-Marquardt algorithm (LMA), one of the most widely used optimization algorithms that can be applied to feed-forward neural networks (Marquardt 1963). The derivative function of the RNN networks was the backpropagation through time (Rumelhart and McClelland 1986a), that is a gradient based technique that begins by unfolding the recurrent neural network through time into feed-forward neural networks, so that the training then proceeds in a manner similar to training a feed-forward neural network with classic backpropagation, except that each epoch must run through the observations in sequential order. The competitive layer classifier is implemented using the `softmax` transfer function that gives as output the probability/likelihood of each classification. It ensured all of the output values were between 0 and 1, and that their sum was 1.
The `softmax` function $\varsigma$ used is:

$$\varsigma(\mathbf{q}_i) = \frac{e^{\mathbf{q}_i}}{\sum_{j=1}^{n} e^{\mathbf{q}_j}} \tag{9.1}$$

where the vector $\mathbf{q}$ is the net input to a softmax node, and $n$ is the number of nodes in the softmax layer.

The architecture of the hidden layers of RNNs was chosen after a performance test, in which after 100 runs with varying number of hidden neurons, the best trade-off solutions were selected in terms of minimization of the error and number of iterations needed to converge. It was found that 10 neurons were not the ideal solution, this because 10 is also the number of different states to represent. Furthermore, in the preliminary experiments the authors of the study also found that the pure linear transfer functions for the hidden layers were more effective than the usual sigmoid. They thus chose not to use a bias or set them to zero for the RNN. Due to these choices, when the networks were not active, all activations were zero, but subsequently could be activated by incepting the activation values to the respective neurons in order to start counting from a specific number.

In addition to the main blocks, an associative network was included in the system to initiate the computation of the system and to implement the number manipulation. After the RNNs learned the number sequence, the switch was needed to stop the counting and to redirect the signals to the competitive classifier for the processing of the result.

Figure 9.4 shows the details of the switch/associative layer that, once the two systems had learned to count, allowed them to operate and communicate with each

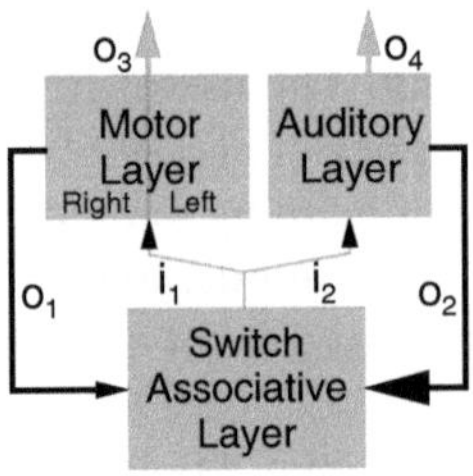


| S= | $\mathbf{i}_1(t)$ | $\mathbf{i}_2(t)$ | $\mathbf{o}_1(t+1)$ | $\mathbf{o}_2(t+1)$ | $\mathbf{o}_3(t+1)$ | $\mathbf{o}_4(t+1)$ |
|---|---|---|---|---|---|---|
| **idle** | 0 | 0 | 0 | 0 | *NaN* | *NaN* |
| **start** | 1 | 1 | $\Sigma \mathbf{i}_1(t) \times \mathbf{h}_1$ | $\Sigma \mathbf{i}_2(t) \times \mathbf{h}_2$ | *NaN* | *NaN* |
| **count** | $\mathbf{o}_1(t)$ | $\mathbf{o}_2(t)$ | $\Sigma \mathbf{i}_1(t) \times \mathbf{h}_1$ | $\Sigma \mathbf{i}_2(t) \times \mathbf{h}_2$ | *NaN* | *NaN* |
| **done** | 0 | 0 | 0 | 0 | $\mathbf{o}_1(t)$ | $\mathbf{o}_2(t)$ |
| **sum** | $\mathbf{o}_1(t)$ | 1 | $\Sigma \mathbf{i}_1(t) \times \mathbf{h}_1$ | $\Sigma \mathbf{i}_2(t) \times \mathbf{h}_2$ | *NaN* | *NaN* |
| **f2w** | *NaN* | $\mathbf{o}_1(t) \times \mathbf{w}_1$ | *NaN* | $\Sigma \mathbf{i}_2(t) \times \mathbf{h}_2$ | *NaN* | *NaN* |
| **w2f** | $\mathbf{o}_2(t) \times \mathbf{w}_2$ | *NaN* | $\Sigma \mathbf{i}_1(t) \times \mathbf{h}_1$ | *NaN* | *NaN* | *NaN* |

× = scalar product; f2w = fingers to words; w2f = words to fingers; NaN preserves the activation at time t (or doesn't activate the units); **w** and **h** are weight vectors, of associative layer and motor/auditory layer respectively.

**Fig. 9.4** Details of the Switch/Associative Layer. The the table on the right summarizes the outputs according to the different states. In practice the layer operates as a recursive feedback with the possibility to start and reset the motor/auditory layers and to derive the activations of one layer from the ones of the other. *Bold lines* indicate a full weighted connection, while *normal lines* are single connections. For simplicity hidden units of the two RNNs of the motor system are represented with one block (Adapted from De La Cruz et al. (2014))

other. In particular, the unit was responsible for starting the counting by initializing all the hidden units to 1, and redirecting the hidden unit activation to the competitive classifier when the counting was finished. Furthermore, this unit was crucial in the development of the acquisition of the ability to add numbers, because it could reset one of the two networks to make it count the new operand, and let the other continue as a buffer memory. Finally, thanks to the associative connections between the two layers (with weights w1 and w2 in Fig. 9.4) there were another two states that allowed inputting a specific number representation starting from another: from fingers to words and vice versa. These states were studied in more detail in the subsequent number manipulation experiments. All states are reported in the table on the left of Fig. 9.4.

As can be seen from the switch/state table in Fig. 9.4 the initial state of all the neurons of the hidden layers were set to 1 in order to start the sequence. Vice versa if the initial state was set to 0, there was no activation because RNNs do not have bias in the hidden layer.

Using the material and methods presented above, the authors of this study first investigated the part of the cognitive system that learned to count. As second step, they built on this by developing the capacity of the associative network to control basic operations like the addition of two operands and derive the number representation of one of the networks from the other (i.e., from fingers to words and vice versa).

In the first experiment which focused on the model's number learning, the main goal of these researchers was to test the ability of the proposed cognitive system to learn numbers by comparing the performance of different ways of training the number knowledge of the robot with:

1. the internal representation (hidden units activation) of a given finger sequence;
2. the Mel-Frequency Cepstral Coefficients (MFCC) coefficients of number words out of sequence;

3. the internal representation of the number words sequence;
4. the internal representation of finger sequences plus the MFCC of number words out of sequence (i.e., learning words while counting);
5. internal representations of the sequences of both fingers and number words together (i.e., learning to count with fingers and words).

To this end, they set up the experiment with the following steps:

i. the motor controller learned the opening of the fingers in a given sequence; in order to later establish a finger counting routine, and create an internal representation for each step in the sequence by means of the hidden units activations;
ii. MFCCs were extracted from number words;
iii. the auditory memory learned the verbal number words in order from 1 to 10 and created an internal representation for each word in the sequence.

From each learning step, relevant data was collected and stored as datasets for the experimentation, these sequences are summarized as follows:

1. Internal representations of the finger sequence: 10 values corresponding to the activation values of the hidden units of motor controller/memory network;
2. MFCCs from number words: 13 values, not as part of a sequence;
3. Internal representations of the words sequence: 10 values from hidden units' activations of auditory memory network;
4. Internal motor representations of the finger sequence and MFCCs: a total of 23 values obtained by merging 1 and 2;
5. Internal motor and auditory representations: a total of 20 inputs obtained by merging 1 and 3.

Data sets were built to model the learning when both fingers and number words were presented together as training input to the cognitive system.

Figure 9.5 shows the activation values of hidden layers of RNNs: Finger sequences on the left and word sequences on the right. The activations of the two RNNs that compose the motor controller/memory network are presented together. Motor activations show lateralization because the network that controls the left hand

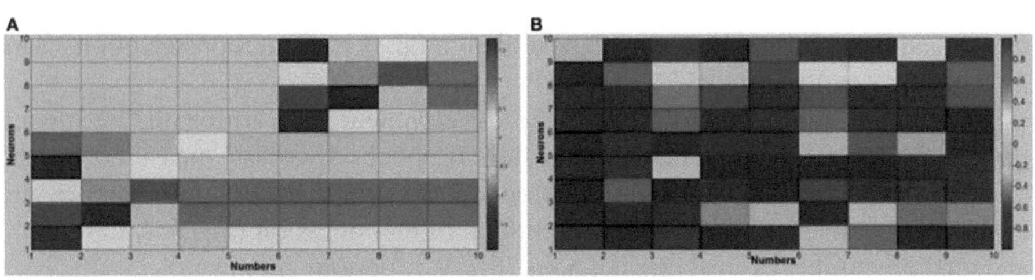


**Fig. 9.5** Activation of the hidden units with the number sequences from 1 to 10. (**a**) RNN trained with finger sequence. (**b**) RNN trained with word sequences

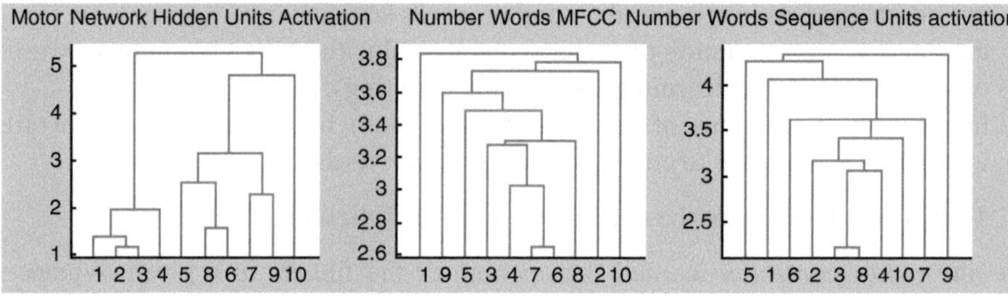


**Fig. 9.6** Optimal leaf-order of the activations of the hidden units

(neurons 6–10) is switched off, furthermore the units from 1 to 5 remain fixed from the number five on, because it was supposed that the right hand was open (reasoning as if the robot was right handed).

In Fig. 9.6, we see a dendogram after the optimal leaf order. It shows how the internal finger representation is more similar to the number sequence, with numbers that were close in the actual sequence closely linked together. On the other hand, the grouping of number words (learned in or out of sequence) is more random, and affects the learning as shown in the classification experiment.

All datasets were used to train the competitive layer classifier to be classified in the ten classes that represent the numbers from 1 to 10. Classification results already after 10 epochs of training are enough for the LMA to converge. After 100 runs for each classification training dataset, the only misclassification observed was for the number three. All the other datasets were good. When fingers and words were presented together, the cognitive system learned numbers quickly and with a very good likelihood, greater than 90 % for all numbers.

Pairwise t-test were also used to evaluate the statistical significance of the results, confirming that all the differences were statistically significant except for the number three, when finger sequences were compared with word sequences, and two when finger sequences only were compared with finger sequences and number words. The "finger sequence and number words", showed that associating the number words with the fingers sequence helped to drastically improve the classification performance without needing to learn number words in a sequence. However, to learn number words in sequence helped to additionally improve the classification performance to the highest likelihood, if internal representations were associated to motor ones.

In order to study in more detail the development of learning in the model, the classification of performance over the 10 epochs was also measured for the competitive layer trained with the different datasets. In this case, performance was evaluated by means of the average likelihood of classification (Fig. 9.7, top graph) and median number of misclassifications (Fig. 9.7, bottom graph)

Looking at the developmental results, it was once again observed that number words learned out of sequence were the less efficient to learn as there were no misclassification only after 10 epochs, and the average likelihood was still low

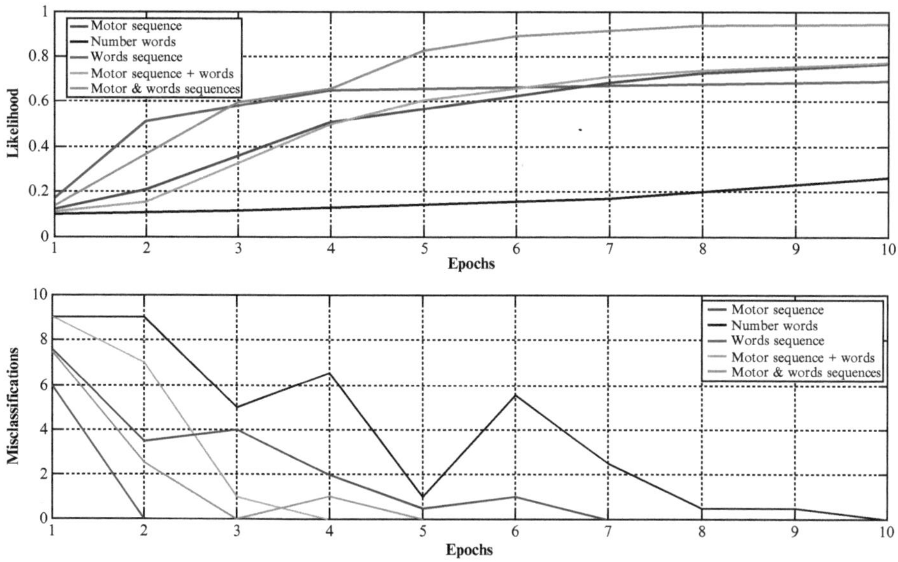


**Fig. 9.7** Development of the model at different epochs of learning: in the *top graph* the average likelihood with number classes, in the *bottom graph* median number misclassification

(0.256) after ten epochs. Conversely, if number words were learned in sequence and internal representations were used as inputs, the learning was faster in terms of the precision of classification (i.e., no errors after just 2 epochs) but the maximum average likelihood, that converged at 0.688, was not as strong as when the learning also involved fingers. Indeed, the finger sequence reached a higher average likelihood (0.765), but best results were obtained when internal representation of words and fingers were used together as input, in fact, the average median likelihood was 0.94 just after 8 epochs.

Once the number sequences were learned, the authors of this study, decided to investigate the possibility of having the model build up the ability to manipulate numbers with the development of the switch-associative network. They did this by extending the capabilities of the associative network from the simple start and stop, to its transferring and mapping, to the basic operation of addition. By transferring, they intended the new mapping of the network's representation derived from the number counted by the other network, when the robot heard the number word "three", to the correlated finger representation. This can be considered, in a sense, an associative mapping between internal representations. This is implemented by activating a weighted connection between the two networks, which can be learned by applying the LMA to the two-layer network that comprises the hidden units of both networks.

The operation of addition can be seen, according to the authors of this study, as a direct development of the concurrent learning of the two recurrent units (motor and auditory). If one of the two does the actual counting of the operands, the other can be used as a buffer memory to add the result, when it is done, the final number can then be transferred from the buffer to the other unit and then inputted to the final processor (the classifier in the system).

As an example they consider 2 + 2, and describe the steps their model takes to compute the result:

1. The first operand is heard by the auditory system and both networks count until the corresponding activation of number 2 is reached. This step corresponds to the states of the associative network.
2. The sum operator is recognized so the auditory network is reset, while the first operand remains stored in the motor memory.
3. The second operand is heard, both networks restart to count as in step 1, until the auditory network reaches the activation corresponding to the number 2. In the meantime, the motor network reaches the activation of the number 4.
4. After the auditory network stops, the associative network recognizes that the work is done so the total (4) is incepted from the fingers network to the auditory network thanks to the associative connection.
5. Finally, the output of the resulting number (4) is produced for final processing (in this case the classifier).

The results obtained in the number learning and number manipulation experiments with the iCub child-like robotic platform, briefly described above, show that learning the number words in sequence along with finger configurations helps the fast building of the initial representation of number in the robot. Number knowledge is instead, not as efficiently developed when number words are learned out of sequence without finger counting. Furthermore, the internal representations of the finger configurations themselves, developed by the robot as a result of the experiments, sustain the execution of basic arithmetic operations, something consistent with evidence coming from developmental research with children. For further details on the cognitive architecture of the developmental cognitive robot model discussed above and the results of the experiments, please see De La Cruz et al. (2014).

The work mentioned above does not intend to suggest that just learning the counting sequence from one to ten, is enough for children (or the particular robot model discussed), to understand number concepts. What it does suggest is something in keeping with what has been proposed by Sarnecka and Carey (2008), that it is the repeated experience using the number word sequence when counting sets of things that might very well be a driving force in the development of numerical understanding .

It has been argued in the literature that the use of fingers does not necessarily precede the use of language in the acquisition of a symbolic numerical system (e.g. Nicoladis et al. 2010). What many children seem to be doing initially, is learning small number word sequences by rote, and later, associations between these small number words and objects in the world (first among which, their readily available fingers). Later on in development, with the child's early schooling experience, this mapping will also include written representations (or numerals). These written representations, eventually take on the meaning of the spoken number word (Fuson 1988). It is this kind of associative multi-modal learning that in a sense is reproduced in the robotic model discussed above.

A number of proposals have been put forth here to address how children learn numbers, number words, the concepts related to them, and the different meanings number words can have. They reflect the results of studies from the domains of developmental psychology as well as theoretical linguistics, with recent work showing an increasing reciprocal awareness of the contributions made by each. This in turn, has lead to more collaborative efforts in addressing what continue to be fascinating yet vexing questions. Very recent work on the development of number cognition in other areas, such as those working within an embodied cognition paradigm, using computational and robotic modeling in particular, have also been discussed in an effort to show how this body of research might also be useful in shedding additional light on these issues. The maturity of the domains involved in investigating the semantic question of number word meaning, the growing knowledge on all the brain components involved in number representation, the experience being gathered with the models here described, are reasons to be persuaded that in the not too distant future, a neurosemantic approach might be feasible in tackling this question as well.

## 9.3 What Next?

How our mind constructs meaning is an age-old question. How our *brain* constructs meaning, and in particular, linguistic meaning, is a relatively recent one. The advent of sophisticated neuroscientific research methods and instruments, has provided the possibility of not only asking this question, but of realistically hoping to find the answers. In the different fields of study briefly discussed in this book, that have each endeavored to find an answer to how meaning is acquired and used in language, such as philosophy, developmental psychology, linguistics, and last but not least, computational modeling, one common thread can be found, or perhaps it would be more accurate to say, not found, in varying degrees. This missing thread, is a serious reflection and/or realistic hypothesis on how the particular phenomenon discussed might be based on brain structure and function, as is they are today understood.

The neurosemantic enterprise, while still in its infancy so to speak, can be considered as a child of our neuroscientific era. It tackles questions that have vexed scholars throughout the centuries that were committed to understanding the relationship between language and the mind, but it does so, not only in a brain-inspired but *brain-informed* manner. Only in the last couple of years has it become feasible to construct neurocomputational models that reflect brain representational mechanisms, as those described in Chap. 3, with components corresponding, at least in part, to relevant cortical and subcortical areas of the brain. There is no going back.

Future progress in the field will have to further address the neuromechanics of the brain and provide explanations on the why and how different areas of the brain (as well as the body through sensorimotor processes) are involved in the acquisition of linguistic meaning and the representation and processing of that meaning (Pulvermüller 2010). Of course, a number of challenges are already present and even more await.

One challenge is that of further grounding word meanings in sensory processes and neuroscientifically confirmed neuronal learning strategies employed in the brain, such as coincidental or correlational learning, using neurocomputational modeling, something being tackled by the work done by the authors of this book, as well as others, and discussed throughout the book. Attempts at grounding linguistic theories, and in particular, those regarding syntax in brain processes, and in particular, in the combinatorial capacities of neurons, is another of the significant challenges currently engaged in by researchers in the neurosemantics domain, as discussed in Sect. 9.1. Another yet, is that of grounding in brain function, the learning of abstract word meanings, such as those pertaining to color, number and morality, also addressed in this book in Chaps. 7, 8, and in Sect. 9.2. A particular challenge for the neurosemantics paradigm, that is considered of utmost importance by the authors of this book, is that of implementing a life span approach, or one that takes into consideration how brain organization and function change throughout the life span, incorporating data on linguistic and conceptual processing at different stages of the development of the brain and of the body (Smith 2013; Wellsby and Pexman 2014). A further challenge, is presented by other types of developments, in the sense of research fields that are themselves also changing and evolving, like those currently using artificial agents in the study of different kinds of cognition. The field of developmental cognitive robotics, for example, working within the embodied cognition paradigm, presents a challenge for the neurosemantics enterprise as well. In fact, the design and creation of brain inspired, brain based neural architectures produced by research in the neurosemantics domain, would find their ideal implementation and verification as controllers in these agents.

Though much remains to be done, work in many of these directions is already under way. The future for a consolidated and mature neurosemantics looks promising.

## References

Adams, S., Wennekers, T., Cangelosi, A., Garagnani, M., & Pulvermüeller, F. (2014). Learning visual-motor cell assemblies for the iCub robot using a neuroanatomically grounded neural network. In *Proceedings of the IEEE Symposium on Computational Intelligence, Cognitive Algorithms, Mind and Brain*, Orlando (pp. 1–8).

Ahmad, K., Casey, M., & Bale, T. (2002). Connectionist simulation of quantification skills. *Connection Science, 14*, 1739–1754.

Andres, M., Luca, S. D., & Pesenti, M. (2008). Finger-counting: The missing tool? *Behavioral and Brain Science, 31*, 642–644.

Asada, M., Hosoda, K., Kuniyoshi, Y., Ishiguro, H., Inui, T., Yoshikawa, Y., Ogino, M., & Yoshida, C. (2009). Cognitive developmental robotics: A survey. *IEEE Transactions on Autonomous Mental Development, 1*, 12–34.

Butterworth, B. (2005). The development of arithmetic abilities. *Journal of Child Psychology and Psychiatry, 46*, 3–18.

Cangelosi, A., & Schlesinger, M. (2015). *Developmental robotics; from babies to robots*. Cambridge: Cambridge University Press.

Cangelosi, A., Greco, A., & Harnad, S. (2001). Symbol grounding and the symbolic theft hypothesis. In A. Cangelosi & D. Parisi (Eds.), *Simulating the evolution of language* (pp. 3–20). Berlin: Springer.

Cantrell, L., & Smith, L. B. (2013). Open questions and a proposal: A critical review of the evidence on infant numerical abilities. *Cognition, 128*, 331–352.

Carey, S. (2001). Cognitive foundations of arithmetic: Evolution and ontogenesis. *Minds and Language, 16*, 37–55.

Carey, S. (2004). Bootstrapping and the origins of concepts. *Daedalus, 133*(1), 59–68.

Chomsky, N. (1964). *Current issues in linguistic theory*. The Hague: Mouton & Co.

Chomsky, N. (1966). *Cartesian linguistics: A chapter in the history of rationalist thought*. New York: Harper and Row Pub. Inc.

Corre, M. L., & Carey, S. (2007). One, two, three, four, nothing more: An investigation of the conceptual sources of the verbal counting principles. *Cognition, 105*, 395–438.

Costa, A. J., Silva, J. B. L., Chagas, P. P., Krinzinger, H., Lonneman, J., Willmes, K., Wood, G., & Haase, V. G. (2011). A hand full of numbers: A role for offloading in arithmetics learning? *Frontiers in Psychology, 2*, 368.

De La Cruz, V. M., Di Nuovo, A., Di Nuovo, S., & Cangelosi, A. (2014). Making fingers and words count in a cognitive robot. *Frontiers in Behavioral Neuroscience, 8*, 13.

Dehaene, S. (1997). *The number sense: How the mind creates mathematics*. Oxford: Oxford University Press.

Di Nuovo, A., De La Cruz, V. M., Cangelosi, A. (2014a). Grounding fingers, words and numbers in a cognitive developmental robot. In *Proceedings of the IEEE Symposium on Computational Intelligence, Cognitive Algorithms, Mind and Brain*, Orlando (pp. 9–15).

Di Nuovo, A., De La Cruz, V. M., Cangelosi, A., & Di Nuovo, S. (2014b). The icub learns numbers: An embodied cognition study. In *Proceedings of the IEEE World Congress on Computational Intelligence*. Beijing.

Domahs, F., Moeller, K., Huber, S., Willmes, K., & Nuerk, H. C. (2010). Embodied numerosity: Implicit hand-based representations influence symbolic number processing across cultures. *Cognition, 116*, 251–266.

Elman, J. L. (1990). Finding structure in time. *Cognitive Science, 14*, 179–221.

Elman, J. L. (1991). Distributed representations, simple recurrent networks, and grammatical structure. *Machine Learning, 7*, 195–225.

Fischer, M. H., Kaufmann, L., Domahs, F. (2012). Finger counting and numerical cognition. *Frontiers in Psychology, 3*, 108.

Fuson, K. C. (1988). *Children's counting and concepts of number*. Berlin: Springer.

Garagnani, M., Wennekers, T., & Pulvermüller, F. (2008). A neuroanatomically-grounded hebbian learning model of attention-language interactions in the human brain. *European Journal of Neuroscience, 27*, 492–513.

Gelman, R., & Gallistel, C. R. (1978). *The child's understanding of number*. Cambridge: Harvard University Press.

Gennari, S. P., & MacDonald, M. C. (2005). Acquisition of negation and quantification: Insights from adult production and comprehension. *Laterza, 13*, 125–168.

Geurts, B. (2003). Quantifying kids. *Language Acquisition, 11*, 197–218.

Gilbert, C. D., Wiesel, T. N. (1983). Clustered intrinsic connections in cat visual cortex. *Journal of Neuroscience, 3*, 1116–1133.

Gracia-Bafalluy, M., Noël, M. P. (2008). Does finger training increase young children's numerical performance? *Cortex, 44*, 368–375.

Harnad, S. (1990). Symbol grounding problem. *Physica D, 42*, 335–346.

Horn, L. R. (1972). On the semantic properties of the logical operators in English. PhD thesis, Indiana University.

Huang, Y. T., Snedeker, J., & Spelke, E. (2012). What exactly do numbers mean? *Language Learning and Development, 0*, 1–26.

Hurewitz, F., Papafragou, A., Gleitman, L., & Gelman, R. (2006). Asymmetries in the acquisition of numbers and quantifiers. *Language Learning and Development, 2*, 76–97.

Kaplan, D. M. (2011). Explanation and description in computational neuroscience. *Synthese, 183*, 339–373.

Kaplan, D. M., & Craver, C. F. (2011). Towards a mechanistic philosophy of neuroscience. In S. French & J. Saatsi (Eds.), *Continuum companion to the philosophy of science* (pp. 268–292). London: Continuum.

Karmiloff-Smith, A. (1992). *Beyond modularity: A developmental perspective on cognitive science*. Cambridge: MIT.

Lafay, A., Thevenot, C., Castel, C., & Fayol, M. (2013). The role of fingers in number processing in young children. *Frontiers in Psychology, 4*, 488.

Levinson, S. C. (2000). *Presumptive meanings*. Cambridge: MIT.

Ma, Q., & Hirai, Y. (1989). Modeling the acquisition of counting with an associative network. *Biological Cybernetics, 61*, 271–278.

Marquardt, D. W. (1963). An algorithm for least squares estimation of non-linear parameters. *SIAM Journal, 11*, 431–441.

Metta, G., Natale, L., Nori, F., Sandini, G., Vernon, D., Fadiga, L., von Hofsten, C., Rosander, K., Lopes, M., Santos-Victor, J., Bernanrdino, A., & Montesano, L. (2010). The iCub humanoid robot: An open-systems platform for research in cognitive development. *Neural Networks, 23*, 1125–1134.

Musolino, J. (2004). The semantics and acquisition of number words: Integrating linguistic and developmental perspectives. *Cognition, 93*, 1–41.

Musolino, J. (2009). The logical syntax of number words: Theory, acquisition and processing. *Cognition, 111*, 24–45.

Nicoladis, E., Pika, S., & Marentette, P. (2010). Are number gestures easier than number words for pre-schoolers? *Cognitive Development, 25*, 247–261. doi:10.1016/j.cogdev.2010.04.001.

Papafragou, A., & Musolino, J. (2003). Scalar implicatures: Experiments at the semantics-pragmatics interface. *Cognition, 86*, 253–282.

Pearlmutter, B. A. (1989). Learning state space trajectories in recurrent neural networks. *Neural Computation, 1*, 263–269.

Pulvermüller, F. (1999). Words in the brain's language. *Behavioral and Brain Science, 22*, 253–336.

Pulvermüller, F. (2002a). A brain perspective on language mechanisms: From discrete neuronal ensembles to serial order. *Progress in Neurobiology, 67*, 85–111.

Pulvermüller, F. (2002b). *The neuroscience of language: On brain circuits of words and serial order*. Cambridge: Cambridge University Press.

Pulvermüller, F. (2010). Brain embodiment of syntax and grammar: Discrete combinatorial mechanisms spelt out in neuronal circuits. *Brain and Language, 112*, 167–179.

Pulvermüller, F. (2013). Semantic embodiment, disembodiment or misembodiment? In search of meaning in modules and neuron circuits. *Brain and Language, 127*, 86–103.

Pulvermüller, F., & Knoblauch, A. (2009). Discrete combinatorial circuits emerging in neural networks: A mechanism for rules of grammar in the human brain? *Neural Networks, 22*, 161–172.

Pulvermüller, F., Härle, M., & Hummel, F. (2001). Walking or talking? Behavioral and neurophysiological correlates of action verb processing. *Brain and Language, 78*, 143–168.

Rajapakse, R. K., Cangelosi, A., Coventry, K. R., Newstead, S., & Bacon, A. (2005). Connectionist modeling of linguistic quantifiers. In *Artificial neural networks; formal models and their applications*. Lecture Notes in Computer Science (Vol. 3697, pp. 679–684). Berlin: Springer.

Rodriguez, P., Wiles, J., & Elman, J. L. (1999). Language acquisition in the absence of explicit negative evidence: How important is starting small? *Cognition, 72*, 67–109.

Rucinski, M., Cangelosi, A., & Belpaeme, T. (2012). Robotic model of the contribution of gesture to learning to count. In *International conference on epigenetic robotics*. San Diego, CA, USA.

Rumelhart, D. E., & McClelland, J. L. (1986a). On learning of past tenses od english verbs. In D. E. Rumelhart & J. L. McClelland (Eds.), *Parallel distributed processing: Explorations in the microstructure of cognition* (pp. 216–271). Cambridge: MIT.

Sarnecka, B. W., & Carey, S. (2008). How counting represents number: What children must learn and when they learn it. *Cognition, 108*, 662–674.

Smith, L. B. (2013). It's all connected: Pathways in visual object recognition and early noun learning. *American Psychologist, 68*, 618–629.

Spelke, E., & Tsivkin, S. (2001). Initial knowledge and conceptual change: Space and number. In M. Bowerman & S. Levinson (Eds.), *Language acquisition and conceptual development*. Cambridge: Cambridge University Press.

Syrett, K., Musolino, J., & Gelman, R. (2012). How can syntax support number word acquisition? *Language Learning and Development, 8*, 146–176.

Tikhanoff, V., Cangelosi, A., Fitzpatrick, P., Metta, G., Natale, L., & Nori, F. (2011). An open-source simulator for cognitive robotics research: The prototype of the icub humanoid robot simulator. In *Proceedings of IEEE workshop on performance metrics for intelligent systems workshop*. Gaithersburg, MD, USA.

Tschentscher, N., Hauk, O., Fischer, M. H., & Pulvermüller, F. (2012). You can count on the motor cortex: finger counting habits modulate motor cortex activation evoked by numbers. *NeuroImage, 59*, 3139–3148.

Wellsby, M., & Pexman, P. M. (2014). Developing embodied cognition: Insights from children's concepts and language processing. *Frontiers in Psychology, 5*, 506.

Witzel, C., & Gegenfurtner, K. R. (2011). Is there a lateralized category effect for color? *Journal of Vision, 11*, 1–25.

Wynn, K. (1992). Children's acquisition of number words and the counting system. *Cognitive Psychology, 24*, 220–251.

Wynn, K., & Bloom, P. (1997). Linguistic cues in the acquisition of number words. *Journal of Child Language, 51*, 167–194.

# Index





A. Plebe, V.M. De La Cruz, *Neurosemantics*,
Studies in Brain and Mind 10, DOI 10.1007/978-3-319-28552-8

MIX
Papier aus verantwortungsvollen Quellen
Paper from responsible sources
FSC® C105338

If you have any concerns about our products,
you can contact us on
**ProductSafety@springernature.com**

In case Publisher is established outside the EU,
the EU authorized representative is:
**Springer Nature Customer Service Center GmbH**
**Europaplatz 3, 69115 Heidelberg, Germany**

Printed by Libri Plureos GmbH
in Hamburg, Germany